Hyphenated Techniques in Polymer Characterization

ACS SYMPOSIUM SERIES **581**

Hyphenated Techniques in Polymer Characterization

Thermal–Spectroscopic and Other Methods

Theodore Provder, EDITOR
The Glidden Company
(Member of ICI Paints)

Marek W. Urban, EDITOR
North Dakota State University

Howard G. Barth, EDITOR
DuPont

Developed from a symposium sponsored
by the Division of Polymeric Materials: Science and Engineering
and the Division of Analytical Chemistry,
at the 206th National Meeting
of the American Chemical Society,
Chicago, Illinois,
August 22–27, 1993

American Chemical Society, Washington, DC 1994

Library of Congress Cataloging-in-Publication Data

Hyphenated techniques in polymer characterization: thermal–spectroscopic and other methods / Theodore Provder, editor, Marek W. Urban, editor, Howard Barth, editor

    p.   cm.—(ACS symposium series, ISSN 0097–6156; 581)

"Developed from a symposium sponsored by the Division of Polymeric Materials: Science and Engineering, Inc., and the Division of Analytical Chemistry, at the 206th National Meeting of the American Chemical Society, Chicago, Illinois, August 22–27, 1993."

Includes bibliographical references and indexes.

ISBN 0–8412–3057–9

    1. Polymers—Analysis—Congresses.   2. Thermal analysis—Congresses. 3. Spectrum analysis—Congresses.

    I. Provder, Theodore, 1939– . II. Urban, Marek W., 1953– . III. Barth, Howard G. IV. American Chemical Society. Division of Polymeric Materials: Science and Engineering, Inc. V. American Chemical Society. Division of Analytical Chemistry. VI. American Chemical Society. Meeting (206th: 1993: Chicago, Ill.) VII. Series.

QD139.P6H96   1994
668.9—dc20   94–38926
                                                    CIP

The paper used in this publication meets the minimum requirements of American National Standard for Information Sciences—Permanence of Paper for Printed Library Materials, ANSI Z39.48–1984.   ∞

PRINTED IN THE UNITED STATES OF AMERICA

Foreword

THE ACS SYMPOSIUM SERIES was first published in 1974 to provide a mechanism for publishing symposia quickly in book form. The purpose of this series is to publish comprehensive books developed from symposia, which are usually "snapshots in time" of the current research being done on a topic, plus some review material on the topic. For this reason, it is necessary that the papers be published as quickly as possible.

Before a symposium-based book is put under contract, the proposed table of contents is reviewed for appropriateness to the topic and for comprehensiveness of the collection. Some papers are excluded at this point, and others are added to round out the scope of the volume. In addition, a draft of each paper is peer-reviewed prior to final acceptance or rejection. This anonymous review process is supervised by the organizer(s) of the symposium, who become the editor(s) of the book. The authors then revise their papers according to the recommendations of both the reviewers and the editors, prepare camera-ready copy, and submit the final papers to the editors, who check that all necessary revisions have been made.

As a rule, only original research papers and original review papers are included in the volumes. Verbatim reproductions of previously published papers are not accepted.

M. Joan Comstock
Series Editor

Contents

INDEXES

Preface

THE CURRENT TECHNOLOGICAL DIRECTIONS in polymer-related industries in the 1990s have been driven and shaped by the operative business and societal forces. These forces strongly affect and influence the product development process, which is no longer a sequential process from research and development (R&D) to product introduction into the marketplace. This process of necessity has become highly nonlinear, nonsequential, and iterative in order to shorten the time from product development to market introduction.

The current economic and business climates are causing a revolution in the manner that companies function and operate. The "right-sizing" of companies and the resulting limited available resources are forcing companies to focus on their core business and technological competencies. This focusing has produced a more directed approach to product development. Parallel-path approaches involve concurrent and iterative activities among the R&D, process scale-up, plant manufacturing, quality assurance, and marketing functions. This focused approach to the product development process is affected by a variety of constraints such as safety, health, the environment, waste reduction, energy conservation, product quality, improved product–process–customer economics, and the need to satisfy–delight the customer.

The polymer-related industries must develop quality products focused to specific market needs. They must foster a climate of continuous technological innovation in an environment of efficient resource utilization and cost-effectiveness. The polymer science and technology required to meet product and market needs are generating complex polymer systems, which often consist of blends or composites of a variety of materials. Consequently, analysis based on a single measurement or an average property is no longer adequate to effectively characterize most polymeric materials. Thus, either a combination of techniques or multidimensional analytical approaches are needed to develop the distributive properties of polymers to establish structure–property–processing relationships and to provide fundamental information and insight into the nature of polymers, polymerization mechanisms, and end-use performance. These business and technological driving forces naturally affect materials characterization methodology, development, and application and spawn a need for higher quality information and shortened analysis times both for polymer characterization and for process monitoring and control. To meet these needs, researchers and instrument manufacturers and material characterization investigators have taken advantage of the improvements continuously being made in digital electronics and computer technology to produce improved sensors and highly automated instruments. These advances also have resulted in an increase in the development of hyphenated characterization techniques and their application to materials characterization problems.

This book covers some of the significant advances in hyphenated techniques in polymer characterization with a focus on thermal–spectroscopic techniques and other methods. This book is organized into two sections. The first section focuses on general considerations concerning hyphenated characterization techniques. The second section focuses on coupled thermal techniques and coupled-thermal–spectroscopic techniques. We hope that this book will encourage and catalyze additional activity in hyphenated characterization method development and the application to polymer characterization.

Acknowledgments

We are grateful to the authors for their effective oral and written communications and to the reviewers for their critiques and constructive comments. The ACS Divisions of Polymeric Materials: Science and Engineering and Analytical Chemistry and the Petroleum Research Fund of the ACS are gratefully acknowledged for the financial support of the symposium.

THEODORE PROVDER
The Glidden Company
(Member of ICI Paints)
16651 Sprague Road
Strongsville, OH 44136

MAREK W. URBAN
Department of Polymers and Coatings
North Dakota State University
Fargo, ND 58105

HOWARD G. BARTH
DuPont
Central Research and Development
P.O. Box 80228
Wilmington, DE 19880

September 19, 1994

General Considerations

Chapter 1

Synergism of Hyphenated Techniques in Polymer Analysis

Marek W. Urban

Department of Polymers and Coatings, North Dakota State University, Fargo, ND 58105

The objective of this chapter is to provide an overview of the currently available multi-functional techniques utilized in polymer analysis and identify those features of hyphenated methods that may play a key role in the future polymer analysis. The discussion centers around synergistic aspects of hyphenated techniques, their potential, error analysis, and utilization in multi-dimensional experiments.

While one carefully designed and executed experiment may furnish answers to many questions giving one level of understanding, additional experiments will provide a more clear picture of molecular structures or morphologies involved, thus further enhancing our knowledge on a given problem. Although conducting multiple experiments is an appealing offer, in a typical single experiment at least one specimen of a given sample is required. If one specimen however can be used in multiple experiments, the output of the experiments will provide a wealth of important complementary information. Furthermore, if the experiments are conducted in a sequence one after another, and the output data is processed synergeticaly, they will not only enhance the level of understanding on a given problem, but will also save a considerable amount of usually expensive analytical time. Going to extremes, if all known analytical tools were employed in the analysis of one specimen, a sample of interest would be characterized within a short timeframe and the level of information would have at least as many dimensions as the number of analytical methods involved.

In view of the above considerations the main motivation behind developments of new characterization methods, ultimately leading to advances in analytical technologies and sciences, is to: 1) develop new techniques for qualitative and quantitative analyses and demonstrate their selectivity and sensitivity; 2) develop multi-functional methods allowing simultaneous analysis of a specimen; 3) develop automated multi-functional techniques; and 4) develop on-line automated quality control techniques. Although one could argue that these analytical areas are relatively independent, history has shown that an interplay of new instrumentation ideas and technological advances resulted in the development of new techniques, followed by multi-functional approaches.

0097–6156/94/0581–0002$08.00/0
© 1994 American Chemical Society

Having available a spectrum of the characterization tools, such as one presented in Table A, an analyst is faced with a decision as to what method(s) to choose that will not only provide new insights into the structures and properties of materials, but to make sure that the information obtained from the analysis is fed back into the materials' designing process, thus allowing the designers to improve their efforts. These are the designing process and characterization methodologies that account for continuing progress in sciences, ultimately leading to developments of new technologies.

In the mid 70's the development of many new materials and technologies engaged all those involved in analytical aspects of materials chemistry to search for better, and develop more sensitive, more precise, and more accurate methods of characterization. While the development of individual, highly sensitive instruments was one aspect of the search, a combination of two or more methodologies in a simultaneous analysis of polymeric materials was another expanding avenue. As a result of these activities, enhancement of sensitivity and selectivity of already existing analytical methods, developments of various permutations of techniques, and the development of entirely new approaches to characterization occurred. At that time, however, the primary analytical tools utilized in chemical analysis were gas chromatography - mass spectrometry (GC - MS), liquid chromatography - mass spectrometry (LC - MS), gas/liquid chromatography (GC - LC), liquid chromatography - infrared spectroscopy (LC - IR), liquid chromatography - nuclear magnetic resonance (LC - NMR), and many at that time home-made permutations of IR-UV-NMR-MS. Following Hirschfeld, (1) "...rising tide of alphabet soup threatens to drown us..." and the hyphen seems to be a common feature of all techniques involved. Realizing that combining more instruments in the analysis of one specimen may open a Padora's Box of structural and quantitative features, the idea expanded very rapidly. Today, there are numerous methods that utilize different physical and chemical principles of the analysis of materials, and polymers in particular.

ANALYTICAL SYNERGISM

When two or more analytical techniques are tied together through appropriate interfaces, new dimension is added to the experiment, new data is obtained, and new difficulties arise. Perhaps one of the complications that is most common when combining different techniques is determination of the hyphenated instrument resolution. Regardless whether this is a concentration limit, spectral resolution, or other limiting features, it is essential to realize that the measured quantity of a hyphenated instrument will be the same as that of individual instruments. As an example let us consider a combined GC-FT-IR instrument. The main reason for "hyphenating" two instruments is that after a specimen is injected into a GC port, it becomes separated in a GC column and therefore, FT-IR detection can be accomplished with a better selectivity because GC allows sample separation prior to FT-IR analysis. One principle to keep in mind is that by combining two or more instruments together one does not increase the actual instrumental resolution or detection limits. These as well as other characteristic features remain the same, but selectivity is enhanced. Although there are circumstances when, depending upon the

Table A. A schematic diagram of various techniques and problems associated with polymer analysis.

Separation Methods
- Gas Chromatography GC
- Liquid Chromatography LC
- Thin Layer Chromatography TLC
- Gel Permeation Chromatography GPC
- Mass Spectrometry MS
- Pyrolysis
- Size Exclusion Chromatography SEC
- Orthogonal Chromatography OC
- Vapor Pressure Osmometry VPO
- Field Flow Fractionation FFF

Spectroscopic Methods
- UV-VIS Spectroscopy
- Fourier Transform Infrared FTIR
  - Internal Reflection ATR
  - External Reflection (R-A)
  - Emission Spectroscopy
  - Diffuse Reflectance DRIFT
  - Infrared Microscopy
  - Photoacoustic Spectroscopy PA
  - Rheo-Photoacoustic Spectroscopy
  - Ellipsometry
- Raman Scattering and Raman Microscopy
- Electron Scanning for Chemical Analysis ESCA
  - Auger Spectroscopy
  - X-ray Photoelectron Spectroscopy XPS
- Circular Dichroism Spectroscopy
- Nuclear Magnetic Resonance NMR
- NMR Imaging
- Fluorescence Spectroscopy
- Phosphorescence Spectroscopy
- Luminescence Spectroscopy
- Light Scattering
- X-ray Diffraction
- Electron Diffraction Microscopy

Thermal - Mechanical Methods
- Differential Scanning Calorimetry DSC
- Thermogravimetric Analysis TGA
- Dynamic Mechanical Thermal Analysis DMTA
- Torsional Braid Analysis TBA
- Dielectric Constant Measurements
- Stress-Strain Measurements

Rheological Methods
- Cone/Plate Viscometry
- Co-Axial Cylinder Viscometry
- Parallel Plate Viscometry
- Tensile or Extensional Viscometry
- Dynamic or Oscillatory Rheometry

Morphological Analysis
- Optical Microscopy
- Phase Contrast Microscopy
- Scanning Electron Microscopy
- Scanning Tunneling Microscopy

Effect of Environment
Solubilization and Thermal History
Effect of Low Molecular Weight Monomers
Conformational Changes and Solvent Effect
Viscosity and Conformational Effects
End-Group Analysis
Free Volume
Copolymer Architecture
Polymer Blends
Effects of Additives
Conformational Analysis
Thermal Properties
Mechanical Properties
Stability of Polymers
- Environmental
- Thermal
- Chemical
- Nuclear
Surface and Interfacial Analysis
Adhesion
Surface Depth Profiling
Corrosion

nature of the species and their detection, both sensitivity, specificity, and selectivity may be simultaneously enhanced, the primary motivation for the development of hyphenated techniques such as GC-MS, LC-GC, and GC-IR was increased specificity.

Results of an analysis conducted on a hyphenated instrument are typically calculated from two or more experimental data sets, each of which carries some uncertainty due to random noise or experimental errors. It is therefore worthwhile determining the ways various uncertainties accumulate in the final output from a hyphenated instrument. For simplicity, let us assume that two in-line instruments measure two quantities x and y which depend upon variables p, q, r for x, and s, t, u for y.

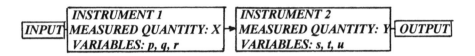

Mathematically, this can be expressed as

$$x = f\,(p,\,q,\,r) \qquad and \qquad y = f\,(s,\,t,\,u) \qquad (1)$$

The uncertainties dx_i and dy_i, that is deviations from the mean values for x and y in the ith measurement will depend upon the size and sign of the corresponding uncertainties dp_i, dq_i, and dr_i for x, and ds_i, dt_i, and du_i for y. That is

$$dx_i = f(dp_i,\,dq_i,\,dr_i) = (x_i - \mu_x)\ and\ dy_i = f(ds_i,\,dt_i,\,du_i) = (y_i - \mu_y) \qquad (2)$$

where μ_x and μ_y are the arithmetic means for an infinite number of measurements. The quantities $x_i - \mu_x$ and $y_i - \mu_y$ are thus the deviations from the mean. The variations in dx and dy as a function of the uncertainties in p, q, r, and s, t, u, can be derived by taking the total differential of eqn. 1, giving

$$dx = (\delta x/\delta p)_{q,r}dp + (\delta x/\delta q)_{p,r}dq + (\delta x/\delta r)_{p,q}dr$$

and $\qquad\qquad\qquad\qquad\qquad\qquad\qquad\qquad\qquad\qquad\qquad (3)$

$$dy = (\delta y/\delta s)_{t,u}ds + (\delta y/\delta t)_{s,u}dt + (\delta y/\delta u)_{s,t}du$$

In an effort to be able to relate the various terms in eqn 3 to standard deviations of x, p, q, r and y, s, t, u as given by the following relationships

$$SD^{\#}_x = \sqrt{\sum_{i=1}^{N}(x_i - x)^2/N}\ \ and\ \ SD_y = \sqrt{\sum_{j=1}^{N}(y_j - y)^2/N} \qquad (4)$$

Note that an abbreviation S.D. is used for standard deviation to distinguish the experimentally determined standard deviation from the theoretical values typically designated as σ.

where N is a total number of samples, it is necessary to square eqn 3, giving

$$dx^2 = \{(\delta x/\delta p)_{q,r}dp + (\delta x/\delta q)_{p,r}dq + (\delta x/\delta r)_{p,q}dr\}^2$$
and
$$dy^2 = \{(\delta y/\delta s)_{t,u}ds + (\delta y/\delta t)_{s,u}dt + (\delta y/\delta u)_{s,t}du\}^2$$

(5)

As a result of this operation, two types of terms from the right-hand side of each equation will be obtained: terms that will be always positive, therefore these terms will never cancel, and cross terms that may be positive or negative. If dp, dq, dr, ds, dt, and du are independent and random, some of the cross terms will be positive and others negative. As a result, the overall summation of the cross terms will approach zero. Therefore, eqn. 5 can be obtained by summation from $i = 1$ to N (where N is a total number of measurements), giving

$$\Sigma(dx_i)^2 = (\delta x/\delta p)^2 \sum_{i=1}^{N}(dp_i)^2 + (\delta x/\delta q)^2 \sum_{i=1}^{N}(dq_i)^2 + (\delta x/\delta r)^2 \sum_{i=1}^{N}(dr_i)^2$$
and
$$\Sigma(dy_i)^2 = (\delta y/\delta s)^2 \sum_{i=1}^{N}(ds_i)^2 + (\delta y/\delta t)^2 \sum_{i=1}^{N}(dt_i)^2 + (\delta y/\delta u)^2 \sum_{i=1}^{N}(du_i)^2$$

(6)

Because both experiments are conducted in sequence one after another, quantities $\sum_{i=1}^{N}(dx_i)^2$ and $\sum_{i=1}^{N}(dy_i)^2$ can be added, giving $\sum_{i=1}^{N}(dz_i)^2$ Dividing both eqns. by N one obtains

$$\Sigma(dz_i)^2/N = \{\sum_{i=1}^{N}(dx_i)^2 + \sum_{i=1}^{N}(dy_i)^2\}/N =$$

(7)

$$(\delta x/\delta p)^2 \sum_{i=1}^{N}(dp_i)^2/N + (\delta x/\delta q)^2 \sum_{i=1}^{N}(dq_i)^2/N + (\delta x/\delta r)^2 \sum_{i=1}^{N}(dr_i)^2/N +$$
$$(\delta y/\delta s)^2 \sum_{i=1}^{N}(ds_i)^2/N + (\delta y/\delta t)^2 \sum_{i=1}^{N}(dt_i)^2/N + (\delta y/\delta u)^2 \sum_{i=1}^{N}(du_i)^2/N$$

From combined eqn. 4, $\sigma = \sqrt{\sum_{i=1}^{N}(z_i - \mu_z)^2/N}$, one obtains

$$(dz_i)^2/N = (z_i - \mu)^2/N = \sigma_z^2$$

(8)

where σ_z^2 is the variance of z (x and y). Similarly, one can represent

$(dp_i)^2/N = \sigma_p^2$; $(dr_i)^2/N = \sigma_r^2$; $(ds_i)^2/N = \sigma_s^2$; $(dt_i)^2/N = \sigma_t^2$;... and so on.

Thus, eqn. 7 can be written in terms of variances as:

(9)

$$\sigma_z^2 = (\delta x/\delta p)^2 \sigma_p^2 + (\delta x/\delta q)^2 \sigma_q^2 + (\delta x/\delta r)^2 \sigma_r^2 +$$
$$(\delta y/\delta s)^2 \sigma_s^2 + (\delta y/\delta t)^2 \sigma_t^2 + (\delta y/\delta u)^2 \sigma_u^2$$

This simple derivation illustrates that when considering a measurement process, the overall variance is a sum of all variables involved in a sequence of experiments and the uncertainties with which each variable is determined propagate to the final output of the experiment.

With this in mind let us go back to the concept of a chemical analysis or for that matter to any chemical process and realize that in order to either perform some type of measurements or improve existing chemical process, or even provide a means for quality control, it is necessary to consider variance. In any chemical process, overall variance will follow the same principles derived in eqn. 9. Typically, the total variance will be a sum of variances coming from various components involved:

$$\sigma^2_{total} = \sum_{j=1}^{n} \sigma^2_{(process + sampling + measurement)} \qquad (10)$$

where $j = 1$ to n, is a total number of operations introducing different variances. Each individual component can be broken down depending upon specific situation. If the task is to obtain the best possible process control, it would be ideal to eliminate sampling variance and measurement variance. While further details concerning statistical approaches related to repeatability-reproducibility (R&R) can be found in the literature,(2 ,3) for the majority of laboratory settings the measurement variance is of a primary concern and can be represented as a sum of variances derived from sample, operator, preparation, and instrumental uncertainties:

$$\sigma^2_{measurement} = \sum_{j=1}^{n} \sigma^2_{(sample + operator + preparation + instrumental)} \qquad (11)$$

Although in this formula it would be ideal to eliminate as many variances as possible, there are inherent limitations. Variances can be minimized, but cannot be eliminated. While instrumental variance is a characteristic feature of instrumentation involved, minimal sample preparation and non-destructive sample participation in the experiment will strongly affect overall variance. In view of these considerations it has been always one of the highest priorities to design a hyphenated instrument that would be non-intrusive and non-destructive to a specimen, yet allowing full characterization.

The concept of synergism to create a complementary scenario for qualitative and quantitative performances of individual techniques is the driving force for many hyphenated techniques. This issue is particularly important in the 90s, as the primary focus is to make each chemical analysis more sensitive and selective, meeting the highest qualitative and quantitative criteria. However, there are inherent complications. Let us consider a copolymer that needs to be characterized. One would introduce a copolymer into an inlet of a hyphenated instrument and force it through a series of interfaced instruments. Each instrument would have a capability of characterizing various properties. Although such proposition seems to be very appealing, in practice, each analytical method may requires different sample treatments. For example, one technique may require sample heating, thus introducing thermal history to the sample, while another may require dissolving a copolymer in a suitable solvent, thus generating morphological changes. In view of the above considerations and realizing that a synergistic approach is usually achieved when several methods chosen for analysis can utilize a common variable, the question becomes what variables can be used. Such common variables can be temperature, oscillating or stationary force, diffusion, viscosity, mobility, magnetic spin, scattering efficiency, fluorescence intensity, solubility, electromagnetic radiation, magnetic fields, to name only a few. For the reasons stated above, many currently existing hyphenated methods utilize only one variable. One, however, would envision that the future approaches may not impose a common requirement for synergism to be achieved if only one variable is common to the techniques of choice. There are already hyphenated techniques where several variables are implemented simultaneously; for example, stress-strain, temperature, and electromagnetic radiation,(4 ,5) or electric current, temperature, and electromagnetic radiation,(6) and a suitable combination of variables appears to be the future of the hyphenated methods. Such approaches will not be possible without computers with fast microprocessors and sensitive detectors that allow fast data acquisition while controlling several variables at the same time.

This chapter is intended not to predict the future, but point out those concepts in multi-disciplinary techniques that may play a considerable role in polymer analysis. Concepts in polymer characterization that may need to be further emphasized in order to advance our knowledge in polymer science can be grouped into several categories. For the purpose of this discussion, the following categories are distinguished:

> Polymers, Copolymer, and Dispersions in Fluidized Forms
> Polymer Structure, Molecular Weight, and Molecular Weight Distribution
>> The Effect of Environment on Polymer Structure
>> Solubilization and the Effect of Thermal History
>> The Effect of Low Molecular Weight Monomers
>> Conformational Changes and Solvent Effect
>> Viscosity and Conformational Effects
>> End-Group Analysis
> Polymers and Copolymers in Solid State
>> Free Volume
>> Copolymer Architecture
>> Polymer Blends
>> Effects of Additives
>> Conformational Analysis
>> Thermal Properties
>> Mechanical Properties
>> Stability of Polymers
>>> Environmental
>>> Thermal
>>> Chemical
>>> Nuclear
> Films and Coatings
>> Surface and Interfacial Analysis
>> Adhesion
>> Surface Depth Profiling
>> Corrosion

Within each of the listed categories one may associate at least two or more experimental techniques that will contribute to its knowledge within each category. In an effort to illustrate these conceptual hyphenation processes Table A was constructed and illustrates which techniques have the highest potential for contributing to the listed categories.

Table A can also serve as a vehicle for creating new hyphenated methods. In this context, the next question comes to mind: How many instruments can be interfaced to form a synergistic outcome in order to obtain desirable information? This issue remains open and it is only a matter of time for more hyphenated techniques to come. Analytical performance of a hyphenated instrument depends primarily on the quality of instruments involved and the efficiency of their interface. This issue is particularly relevant in view of the continuously increasing sensitivity and demands for more accurate quantitative analysis. In an effort to obtain quantitative data each instrument must represent a state-of-the-art unit in its own category that can perform independently, and the interface to another instrument should not disturb overall performance. Among many fields of materials chemistry, polymer characterization, and particularly polymer surfaces and interfaces (7) represent perhaps the most challenging area because "high purity polymer samples or polymer surfaces" do not exist. Molecular weight, molecular weight distribution, surface stoichiometry and morphology, interfacial distribution of various film components, or different stereoisomers make polymers what a main stream chemist would consider "low purity." For that reason, characterization and subsequent understanding of polymeric systems require inputs from many fields: chromatography, spectroscopy, rheology, particle size characterization, surface morphology and spectroscopy, thermal and mechanical analysis, and physical testing. Therefore, many diverse and specialized backgrounds are required in the analysis of polymers. In essence, existing methodologies can be divided into several categories using various criteria. Table B lists not only currently available techniques for separation, spectroscopic analysis, thermal and mechanical analysis, or rheological studies, but also connects various techniques into hyphenated categories.

With this philosophy in mind it is apparent that the spectroscopic group can be hyphenated with almost any other group of the diagram as well as within its own category. These include studies of polymer surfaces by scanning electron microscopy combined with X-ray and Raman microscopy, solving problems ranging from degradation, adhesion, corrosion, and others. Similarly, all categories that are thermal - mechanical can be hyphenated to rheological, morphological, and spectroscopic categories. Other combinations of techniques are given in a diagram presented in Table B.

Multi-dimensional analysis of polymers is a rapidly expanding area of research. While a combination of several separation methods has been used with great success, rather huge instrumental requirements along with a difficulty of data reduction and presentation made multi-dimensional analysis a bit slow. However, new developments in detection systems and computers stimulated developments of such techniques as 2-dimensional LC-SEC chromatography, FFF-Light Scattering, SEC-FT-IR, SEC with electrospray MS detection, small and wide angle X-ray scattering with DCS, multi-dimensional FT-IR, Raman, NMR techniques and others. Ultimately, one would like

Table B. Schematic representation of several existing and potential hyphenated techniques utilized in polymer analysis.

Separation Methods
Gas Chromatography GC
Liquid Chromatography LC
Thin Layer Chromatography TLC
Gel Permeation Chromatography GPC
Mass Spectrometry MS
Pyrolysis
Size Exclusion Chromatography SEC
Orthogonal Chromatography OC
Vapor Pressure Osmometry VPO
Field Flow Fractionation FFF
Spectroscopic Methods
UV-VIS Spectroscopy
Fourier Transform Infrared FTIR
    Internal Reflection ATR
    External Reflection RA
    Emission Spectroscopy
    Diffuse Reflectance DRIFT
    Infrared Microscopy
    Photoacoustic Spectroscopy PA
    Rheo-Photoacoustic Spect. RPA
    Ellipsometry
Raman Scattering and Raman Microscopy
Electron Scanning for Chem. Anal. ESCA
    Auger Spectroscopy
    X-ray Photoelectron Spect. XPS
Circular Dichroism Spectroscopy
Nuclear Magnetic Resonance NMR
NMR Imaging
Fluorescence Spectroscopy
Phosphorescence Spectroscopy
Luminescence Spectroscopy
Light Scattering
X-ray Diffraction
Electron Diffraction Microscopy
Thermal - Mechanical Methods
Differential Scanning Calorimetry DSC
Thermogravimetric Analysis TGA
Dynamic Mechanical Therm. Anal. DMTA
Torsional Braid Analysis TBA
Dielectric Constant Measurements
Stress-Strain Measurements
Rheological Methods
Cone/Plate Viscometry
Co-Axial Cylinder Viscometry
Parallel Plate Viscometry
Tensile or Extensional Viscometry
Dynamic or Oscillatory Rheometry
Morphological Analysis
Optical Microscopy
Phase Contrast Microscopy
Scanning Electron Microscopy
Scanning Tunneling Microscopy
Atomic Force Microscopy

Thermal - Mechanical Methods
Differential Scanning Calorimetry DSC
Thermogravimetric Analysis TGA
Dynamic Mechanical Therm. Anal. DMTA
Torsional Braid Analysis TBA
Dielectric Constant Measurements
Stress-Strain Measurements
Rheological Methods
Cone/Plate Viscometry
Co-Axial Cylinder Viscometry
Parallel Plate Viscometry
Tensile or Extensional Viscometry
Dynamic or Oscillatory Rheometry
Morphological Analysis
Optical Microscopy
Phase Contrast Microscopy
Scanning Electron Microscopy
Scanning Tunneling Microscopy
Atomic Force Microscopy
Spectroscopic Methods
UV-VIS Spectroscopy
Fourier Transform Infrared FTIR
    Internal Reflection ATR
    External Reflection RA
    Emission Spectroscopy
    Diffuse Reflectance DRIFT
    Infrared Microscopy
    Photoacoustic Spectroscopy PA
    Rheo-Photoacoustic Spect. RPA
    Ellipsometry
Raman Scattering and Raman Microscopy
Electron Scanning for Chem. Anal. ESCA
    Auger Spectroscopy
    X-ray Photoelectron Spect. XPS
Circular Dichroism Spectroscopy
Nuclear Magnetic Resonance NMR
NMR Imaging
Fluorescence Spectroscopy
Phosphorescence Spectroscopy
Luminescence Spectroscopy
Light Scattering
X-ray Diffraction
Electron Diffraction Microscopy
Separation Methods
Gas Chromatography GC
Liquid Chromatography LC
Thin Layer Chromatography TLC
Gel Permeation Chromatography GPC
Mass Spectrometry MS
Pyrolysis
Size Exclusion Chromatography SEC
Orthogonal Chromatography OC
Vapor Pressure Osmometry VPO
Field Flow Fractionation FFF

to perform multi-dimensional analysis using hyphenated instruments, and these approaches seem to dominate future developments in polymer analysis.

FINAL REMARKS

The last twenty some years of analytical chemistry, particularly of polymers, has been trying to tackle difficult and complex problems which involved sampling situations of various nature. Although it may seem that the progress was fairly steady, there is a tremendous difference in analytical procedures utilized in the past. Yet, there are still plenty of opportunities to create new combinations of techniques because experimental sciences are profoundly influenced by the development of computerized instruments. This is why sophistication of many future instrumental approaches will be mandated by the use of highly sensitive fast instruments and powerful data acquisition computers to attain insights about fundamental aspects of processes under investigation. A large variety of analytical techniques for polymer characterization have been developed under various conditions, but the ultimate goal is to correlate molecular level structures with macroscopic properties. Therefore, hyphenation of molecular level techniques with other macroscopic tools will give better understanding of macroscopic processes responsible for the behavior of polymers, composites, coatings, and other related hybrid materials. These goals can be accomplished if more emphasis is given to the hyphenated techniques included in Table B.

REFERENCES

1. Hirschfeld, T., Anal.Chem., 1980, Vol. 52, No.2, 297 A.
2. Nelson, W., Applied Life Data Analysis, John Wiley & Sons, New York, 1982.
3. Hognet, R.G., Hartman, D.M., Amer.Cer.Soc.Bull., 1994, Vol. 73(5), 64.
4. Siesler, H.W., *Structure-Property Relations in Polymers; Spectroscopy and Performance*, Adv.Chem.Series, #236, Eds. Urban, M.W., Crever, D.C., American Chemical Society, Washington, DC, 1993
5. Urban, M.W., Goettler, H.J., US Patent, #5,036708, 1991.
6. Dittmar, R.M., Chao, J.L, R.A., Palmer, R., Appl. Spectrosc., 1991, 45, 1104.
7. Urban, M.W., *Vibrational Spectroscopy of Molecules and Macromolecules on Surfaces,* John Wiley & Sons, New York, 1993.

RECEIVED August 25, 1994

Chapter 2

Integrated Intelligent Instruments for Materials Science

S. A. Liebman[1,4], C. Phillips[2], W. Fitzgerald[2], R. A. Pesce-Rodriguez[3],
J. B. Morris[3], and R. A. Fifer[3]

[1]The CECON Group, Inc., 242 North James Street,
Wilmington, DE 19804
[2]CCS Instrument Systems, Inc., Unionville, PA 19375
[3]United States Army Research Laboratory,
Aberdeen Proving Ground, MD 21005

Trace organic analysis in materials science requires instrumentation and methods for applications to natural and synthetic polymers in complex mixtures. Detailed information of polymer microstructure and composition is obtained from integrating sample processing, e.g., analytical pyrolysis, dynamic headspace, and supercritical fluid (SF) technologies, with online chromatographic separations by capillary gas chromatography (GC), SF chromatographic (SFC), and spectral detection systems, such as Fourier transform infrared (FTIR), and/or mass spectrometers (MS). Advanced data handling and interpretation guides are addressed with expert system and neural networks with data that are generated from these highly automated analytical systems. The integrated intelligent instrument (I^3) approach is demonstrated with applications to high performance composites, analysis and processing of consumer products, space research, propellant and demilitarization programs.

Polymer characterization using the tools of thermal analysis, analytical or process chromatography, and spectroscopy has developed primarily within the industrial analytical research and services community for problem-solving and in QA/QC operations. Diverse instruments and methods are needed to

[4]Correspondence address: 91 Pinnacle Road West, Holtwood, PA 17532

0097–6156/94/0581–0012$08.00/0

handle the wide range of thermal sensitivities exhibited by macromolecules important to the polymer processing, food, pharmaceutical, biotechnology, and engineering polymer/composite industries. Over the past two decades highly automated systems were developed to provide specific, often unique information to analysts for both research and routine applications (1-8).

Applied artificial intelligence (AI) tools, such as expert system and neural networks, now permit special expertise to be encoded along with the instrumental tools and methods that generate the data. Early efforts were initiated in development of a prototype expert system network for materials analysis, EXMAT (4), followed by the on-going development of an expert system network for SF technologies, MicroEXMAT (5-9). Use of applied AI in analytical chemistry with commercial software was demonstrated by training neural networks for pattern recognition of chromatographic and spectral data (10-12 and references therein).

This report presents our applied research in polymer characterization that emphasizes the advantages of integrating hardware, software, and interpretive guides to provide precise analytical information of multicomponent materials in an efficient manner. The I^3 approach is outlined in **Figure 1** with a focus on sample processing of materials using automated, interfaced instruments and methods developed with applied AI tools.

**AUTOMATED AND INTERFACED
ANALYTICAL SYSTEMS**
///
**Sample Processing
Monitoring/Sampling
Concentrator/Prep Scale
Separation
Detection/Identification
Data Analysis/Chemometrics
Applied AI**

Figure 1.   Integrated Intelligent Instruments, the I^3 approach.

Experimental

Materials.   Polymer and reference samples include propylene-1-butene copolymers, chlorinated poly(vinyl chloride) (PVC), polybenzobisoxazoles (PBO), hydrogen cyanide (HCN) polymers, ethoxylated alcohol surfactants [polyethylene oxide (10 moles)/mole oleyl alcohol (POE-10) and POE-20 (20 moles/mole)], plasticized cellulose acetate butyrate-based propellants, snack food product and a fluorinated rubber composite product.

Instrumentation and Methods.   Analytical pyrolysis/dynamic headspace studies were conducted with a CDS Model 2000 Pyroprobe/Model 330 Sample Concentrator (CDS Analytical Inc., Oxford, PA 19363) interfaced to either a Varian 3700 GC/FID or a Hewlett-Packard GC/MSD typically equipped with 30 m x 0.53 mm, 1 micron film, DB-5 fused silica capillary columns, programmed from 50 C to 380 C at 15 deg/min. A Pyroprobe-Pyroscan/IR cell (CDS Analytical, Inc., Oxford, PA) was used for in-situ monitoring of PVC degradation products by FTIR (Nicholet, Madison, WI). A direct insertion Pyroprobe (DIP) was interfaced with a Finnigan Model 4500 triple quadrupole MS/MS system for time-resolved pyrolysis DIP-MS in the single quadrupole, electron ionization (EI) mode. Other studies used the integrated pyrolysis/concentrator/GC/FID system (CDS Model EA 600 CDS Analytical Inc., Oxford, PA 19363). FT-Raman data were obtained with a Bomem Hartmann & Braun (Quebec, Canada) DA-8 spectrometer system.

Sample processing with SF extraction (SFE)-SFC or SF reaction (SFR)-SFC modes was conducted for analytical or process treatments using a CCS Model 10000 SFC/GC/FID system or a CCS Model 3100-100 SF Processor (CCS Instrument Systems, Inc., Unionville, PA). A CCS Transcap interface permitted off-line FTIR analysis of SF extractables obtained in the SF liquid carbon dioxide mobile fluid. Analytical SFE runs were typically 540-680 atm (8000-10,000 psi) 10-15 min., 80-100 C, with ca. 10 mg samples in 1.5 mL SF vessels. Process SFE runs were conducted in a 100 mL SFE vessel at 340 atm (5000 psi), 80 C for up to several hours to remove oligomers from small (ca. 5 gm) composite rubber products. Analyses were conducted using an off-line FTIR-microscope assembly (Mattson Instruments, Madison, WI and Spectra-Tech, Stamford, CT).   Food materials (ca. 80 gm) were processed in the benchscale SFE Processor at 540 atm (8000 psi), 40 C for 15 min. with collection of the extractables for subsequent off-line analysis by SFE-SFC/FID or HPLC/UV (high pressure liquid chromatography with ultra-violet detection).

Results and Discussion

Pyrolysis, Chromatographic, and Spectral Configurations.   Pyrolysis FTIR studies of evolved degradation products from polymerics provide rapid, unique information that is useful in formulating fire-retardant materials, as well as understanding basic combustion phenomena, or obtaining microstructural information. Such data are needed for recycling, incineration or environmental emissions compliance, and for correlating polymer composition/microstructure to performance. **Figure 2a** presents a typical high resolution capillary GC pyrogram from the quantitative analysis of a propylene-1-butene copolymer (containing 11.3% 1-butene) using a **pyrolysis/concentrator/GC/FID** system. Correlation of denoted fragment peaks A and B to the 1-butene content was linear in the 5 to 50% range as shown in **Figure 2b.**

Infrared spectra of polymers are also obtained in a rapid screening mode by **pulse pyrolysis-FTIR** using solid samples (ca. 0.1-0.5 mg) that are placed "as is" into the Pyroprobe-Pyroscan-FTIR system for semi-quantitative, qualitative information. The vapor phase IR spectrum in **Figure 3a** is that from a pulse pyrolysis (750 C for 10 sec) of a 100 mg sample of solid poly(styrene). The thermal decomposition of poly(styrene) to its styrene

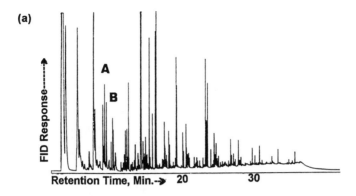

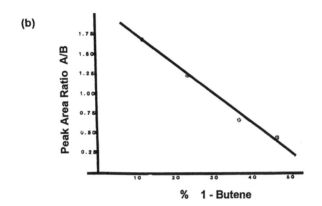

Figure 2. (a) Pyrolysis/Concentrator/GC/FID. Pyrogram from pulse pyrolysis (750 C/ 10 sec) of propylene-1-butene copolymer. **(b)** Peak area ratio A/B versus % 1-butene concentration.

monomer contrasts to that shown for PVC and related vinyls that exhibit side-chain splitting (e.g., HCl) as the initial vapor product , as discussed below.

Figure 3b shows results from a **time-resolved programmed pyrolysis-FTIR** study.   Infrared absorption spectra taken in 45 sec. scans are shown between 4000-2000 cm-1 (x-axis), absorbance units (y-axis), and 100-900 C (z-axis) of chlorinated PVC using the Pyroprobe-Pyroscan-FTIR system. **Figure 3c** shows the gas phase IR spectrum of evolved HCl.   **Figure 3d** presents the vapor pyrolyzate spectra from the time-resolved pyrolysis (60 C /min to 900 C) of an ethylene/vinyl acetate copolymer with IR scans taken every 45 sec. to follow the fragmentation process that releases acetic acid from the vinyl acetate monomeric unit.   These vinyl-substituted polymerics exhibit initial side-chain splitting from the -C-C- backbone, rather than initial fragmentation to monomeric units, as shown by poly(styrene) and its copolymers (3). Hence, diagnostic use is made of the amount of HCl   or

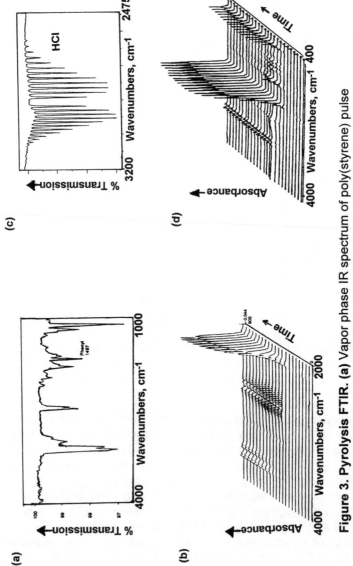

Figure 3. Pyrolysis FTIR. (a) Vapor phase IR spectrum of poly(styrene) pulse pyrolyzed at 750 C/10 sec. **(b) Time-resolved Pyrolysis-FTIR** spectrum of Cl-PVC (60 C/min to 900 C). **(c)** HCl vapor phase FTIR spectrum (3500-2500 cm-1) from Cl-PVC pyrolysis (60 C/min to 900 C in helium). **(d) Time-resolved Pyrolysis-FTIR** spectrum of ethylene-vinyl acetate copolymer (60 C/min to 900 C in helium).

acetic acid products and the temperature at which they are evolved in studies of polymer microstructure, stabilization or degradation kinetics.

Similar monitoring with systems in dynamic headspace configurations also give details of residual monomers, solvents, additives, or study of defect structures with pyrolysis-spectral systems. **Figure 4** shows a series of pyrograms obtained with a **pyrolysis/concentrator/GC/MSD** system of PBO research fibers after process treatments (heated at 600 C, 665 C and 1200 C). Key evolved fragment ions from a pulse pyrolysis at 1000 C/20 sec are identified as benzonitrile (m/z 103) and the isomeric dicyanobenzenes at m/z 128 (benzodicarbonitriles) from comparisons of MS library spectra. Such fragments relate to backbone and/or crosslink structures (e.g., -C=N, -N=C=) that may form during synthesis and/or subsequent heat treatments. Insight into PBO fiber thermal process chemistry and details of composition and microstructure are thus obtained for correlation to final performance. These studies and those relating to cure mechanisms provide basic input to industrial R&D projects in order to develop new materials and/or improve existing processes

In ultra-trace analyses, such as detection of part-per-billion components in a few milligram/microgram sample, use is made of the direct insertion Pyroprobe (DIP) with pulse or programmed thermolysis/pyrolysis and interfacing to a triple quadrupole MS system. **Figure 5a** gives the pyrolysis mass maps (PyMM) from examination of the control (as spun) liquid crystal PBO fiber (7) heated at 120 C/min. to 1000 C. Residue remaining in the Pyroprobe quartz tube coil insert is pulsed pyrolyzed at 1200 C for a 20 sec. interval and is shown in the PyMM at the higher scanset/temperature range.

In the **programmed, time-resolved pyrolysis DIP-MS** run (pyrolyzing at 120 C/min to 1000 C), the three significant fragment ions at m/z 76, 103, and 128 (benzyne, benzonitrile, and the isomeric dicyanobenzenes) comprise only about 3.3 % of the total volatile pyrolyzate as seen (**Figure 5b**) in the reconstructed ion current (RIC). The residue on pulse pyrolysis yields only a small (0.7%), but significant amount of those species. The total ion current (TIC) from the volatile pyrolyzate from both programmed and pulse pyrolysis modes are shown in **Figure 5c**. Such DIP-MS methods are important means to study time-resolved degradation (*1,3*), as well as polymer microstructure/composition; e.g., carbon/graphite fibers and precursors were also investigated following specialty thermal and coating treatments (*7*).

Similar thermal treatments of specially prepared HCN polymers ("multimers") showed their unique high temperature behavior (*13*). Over 90% char remained when a 100 mg sample was heated to 600 C at 20 C/min under nitrogen in TGA runs (*14*) and significant residue was noted in programmed pyrolysis runs to 1200 C under helium or vacuum. These observations relate to the chemistry of refractory, high-charring organics (14), as well as to the role of HCN polymers in advanced space research and prebiotic chemistry (15). Over a dozen instrumental systems/methods were used to screen the complex oligomeric HCN mixture for insight into its composition and component microstructures. The C-N/C-C graphite-like

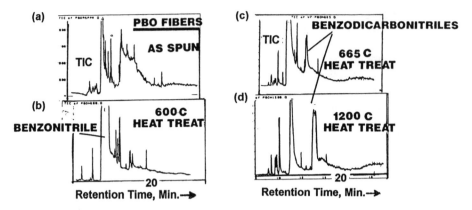

Figure 4. Pyrolysis/Concentrator/GC/MSD. Total Ion Chromatogram **(TIC)** of PBO fibers pulse pyrolyzed at 1000 C for 20 sec. **(a)** as spun **(b)** heat treated 600 C **(c)** Heat treated at 665 C **(d)** Heat treated at 1200 C.

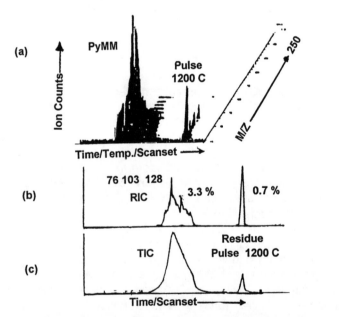

Figure 5. Pyrolysis DIP-MS. (a) Pyroprobe Mass Maps **(PyMM)** of PBO fiber heated at 120 C/ min. to 1000 C, then residue pulsed at 1200 C/ 20 sec. **(b)** Reconstructed Ion Current **(RIC)** for selected ions m/z 76, 103, 128 **(c)** Total Ion Current **(TIC)** of volatile pyrolyzates from sample and from residue.

network resulting from HCN polymerization may serve as a source of new high performance polymerics for use at high temperature.

Nondestructive Infrared and Supercritical Fluid Technologies. In studies that required nondestructive, in-situ analysis with specialty FTIR sampling capabilities, **Figure 6** shows the **microreflectance-FTIR** spectra (1800-1500 cm-1 of a solid propellant at the (a) exterior extruded surface, (b) interior "bulk", (c) end surface and (d) interior "bulk" after exposure to solvent. The peaks labelled RDX and P2 refer to the absorbances due to the energetic component and plasticizer, respectively.

It is shown from these data that the presence of solvent in the formulated material influences the plasticizer (P2) migration. This study of plasticizer and solvent migration in a solid propellant formulation gave insight into the effects of residual solvents, observed "stabilization periods", and general aging of polymer composites. Similar behavior of plasticizers, stabilizers,

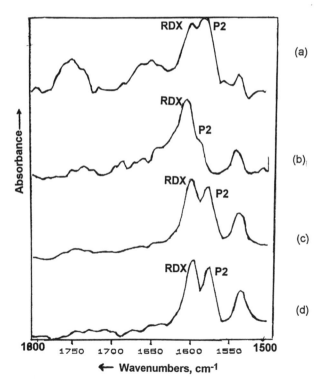

Figure 6. Microreflectance-FTIR. IR Spectra of ethyl acetate solvent in propellant with plasticizer **(a)** exterior extruded surface **(b)** interior "bulk" **(c)** end surface **(d)** interior "bulk" after exposure to solvent.

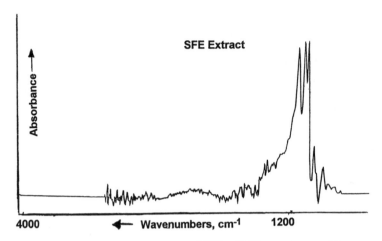

Figure 7. SFE Microreflectance-FTIR. FTIR spectrum of fluoro-polymers extracted from benchscale SFE of rubber composite.

and other additives in consumer products is also of major concern to analysts in industrial R&D laboratories and mandates the use of such methodology.

Polymeric materials that are thermally sensitive require alternative analytical approaches. Again, processing and analysis of the representative sample "as is" can be achieved with varied SF instrumentation and methods. **Figure 7** gives the **off-line SFE microreflectance-FTIR** spectrum of a contaminant removed from a rubber composite product by an SFE in-process treatment at 340 atm (5000 psi) and 100 C over a several hour period. The SFE method used both static (no flow of mobile fluid through the pressurized vessel) and dynamic exposure of the product to a flow of 4-5 mL liquid carbon dioxide mobile fluid at 340 atm (5000 psi). The extracted contaminant was identified as an oligomeric fluorocopolymer based on FTIR spectral library comparisons.   Presence of the contaminant was shown to be detrimental to in-field performance of the product and efficient removal was needed with an environmentally acceptable process.

Other benchscale/ process SFE or SF coating treatments are developing for composite and electronic products (*13*) to replace solvent-based operations with safe, inexpensive liquid carbon dioxide mobile fluid. **Figure 8** shows typical on-line **SFE-SFC** analyses of nonionic surfactants (POE-10 and POE-20) used in development of larger-scale process parameters. Surfactants may be present in degreasing operations commonly used in manufacturing operations for metals, and they may require added removal treatments.   Sensitive cleaning of computer and electronic parts now is conducted using SF technology to ensure the removal of residual surfactants or similar unwanted additives. Pollution prevention is thus achieved with cost-effective processing without organic solvents.

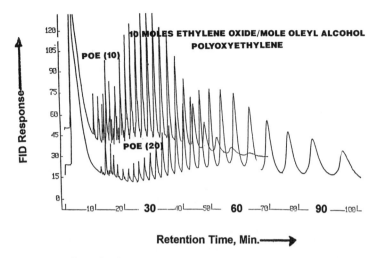

Figure 8. SFE-SFC chromatograms of test nonionic surfactants POE-10 and POE-20 with SFE at 9000 psi/ 20 min. and on-line SFC asymptotic density programming from 0.4 gm/ml to1.0 gm/ml of CO_2 mobile fluid.

Analysis of biopolymers and sensitive polymerics common to the food, biomedical and pharmaceutical industries also benefit from new SF capabilities. **Figure 9** shows the results of **benchscale SFE processing** of a food sample (up to 100 gm) to obtain process engineering scale-up data.
The **off-line SFE-SFC** analysis of the extractables is used to determine the lipid and cholesterol content and to guide the process parameter study. Analytical SFE-SFC instruments and methods also provide such data using on-site analyzers for QA/QC or in-plant operations. More efficient monitoring of high performance materials is achieved, relative to traditional multistep Soxhlet extraction operations, as well as meeting compliance to regulatory environmental and/or quality standards within the industries.

Applied AI in Analytical Chemistry

Availability of commercial software packages (e.g., NeuralWorks II/Plus) permit applications of neural networks to chromatographic, spectral, or nondestructive imaging data (*10,11, 16*). A recent report (*17*) demonstrated the training of neural nets to correlate chemical functional group information to FT-Raman spectral patterns. In that study 212 specific descriptors (e.g., primary aliphatic alcohols, conjugated aliphatic acids, nitroaromatics) and 48 generic functional descriptors (e.g., amines, azides, thiols) were used for correlation to their spectral absorbances. Identification of unknown spectra was demonstrated for compounds that were not in the training set. In addition, the network successfully analyzed spectra of compounds used in

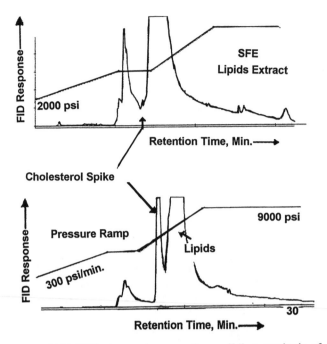

Figure 9. SFE-SFC chromatograms from off-line analysis of
extracted lipids/cholesterol from food product in benchscale
SFE Processor.

the training, even though the spectra had low signal-to-noise ratios. As
applications of neural networks continue for diverse data patterns
(chromatographic, IR, MS, etc.), analysts will benefit from these data-
handling advances in the analysis of complex materials. Prediction of fully
automated sample identification (*18*) using thermal/nonthermal processing
integrated with on-line chromatographic and spectral data appears to
becoming reality as the tools of AI are applied to concerted organic analysis.

Summary and Conclusions

The I^3 approach has been demonstrated with diverse problem-solving
illustrations of polymeric materials characterization.   Detailed, validated
information in the synthesis and processing of complex natural and synthetic
polymerics is imperative, so that the final formulated products meet their in-
field performance requirements. Automated analytical and process
instrumentation with preprogrammed methods and interpretive guides enable
manufacturers to provide high quality products for the global marketplace.

Literature Cited

1. Liebman, S.A., Levy, E.J. In *Polymer Characterization: Spectroscropic, Chromatographic, and Physical Experimental Methods;* Craver, C.D., Ed.; Adv. Chem. Series 203; American Chemical Society, Washington, DC, 1983, pp. 617-624.
2. Courtesy of Applications Lab., CDS Analytical Inc., Oxford, PA 19363.
3. *Pyrolysis and GC in Polymer Analysis;* Liebman, S.A.; Levy, E.J., Eds.; Marcel Dekker, Inc., NY, 1985.
4. Liebman, S.A., Duff, P.A., Schroeder, M.A., Fifer, R.A., Harper, A.M. In *Artificial Intelligence Applications in Chemistry;* Pierce, T.; Hohne, B., Eds.; ACS Sympos. Series 306; American Chemical Society, Washington, DC, 1986, pp. 365-384.
5. Liebman, S.A., Smardzewski, R.R., Sarver, E.W., Reutter, D.J., Snyder, A.P., Levy, E.J., Lurcott, S.M., O'Neill, S. *Proc. Polym. Materials. Sci. Engin.* **1988,** *59,* pp. 621-625.
6. Liebman, S.A., Levy, E.J., Lurcott, S., O'Neill, S., Guthrie, J., Yocklovich, S., *J. Chromatogr. Sci.* **1989,** *27,* pp.118-126.
7. Liebman, S.A., Wasserman, M.B., Snyder, A.P., Pesce-Rodriguez, R.A. Fifer, R.A., Denny, L., Helminiak, T. *Polym.* Preprints **1991,** *32 (2),* pp. 276-278, Special issue; Polymer Tech. Conf., Phila. PA, June,1991.
8. Ryan, T., Yocklovich, S., Lurcott, S., Watkins, Levy, E.J., Liebman, S., Morris, J. *Intelligent Instruments & Computers* **1990,** *8 (3),* pp.109-119.
9. TIMM is an expert system development shell from General Research Corp., Arlington, VA 22209 used for prototype EXMAT and MicroEXMAT.
10. Morris, J.B., Pesce-Rodriguez, R.A., Fifer, R.A., Liebman, S.A., Lurcott, S.M., Levy, E.J., Skiffington, B., Sanders, A., *Intelligent Instruments & Computers* **1991,** *9 (5),* pp.167-175.
11. Brainmaker neural network software is available from California Scientific Software, Sierra Madre, CA 91024. NeuralWare, Inc., Pittsburgh, PA 15276 is supplier of NeuralWorks Explorer and NeuralWorks II/Plus software. ALN software is supplied by General Research Corp., Advanced Technology Div., Arlington, VA 22209.
12. Liebman, S.A., Phillips,C.E., Fitzgerald, W., Levy, E.J. *Proc. Third Internat. Sympos. Field Screening Methods, EPA/EMSL,* Las Vegas, NV, Feb. 1993 and *Second Internat. Sympos.,* Feb. 1991.
13. Liebman, S.A., Levy, E.J., Brame, E.G., Jr. In *Critical Materials and Processes in a Changing World;* Goldberg, Harper, Schroeder, Ibrahim, Eds; SAMPE (Soc. Adv. Materials Process and Engineering): Covina, CA, 1992, Vol. 6; pp. 765-777. Presented at 6th Internat. SAMPE Electronics Conf., Baltimore, MD June,1992.
14. Armistead, P., Mera, A, Naval Research Lab., Washington, DC, personal communication, 1992 and Armistead, P., Sastri, S., Keller, T.M. *Ibid.,* pp. 474-484.
15. Liebman, S.A., Pesce-Rodriguez, R.A., Matthews, C.N. *Adv. Space Res.,* in press, **1994.**
16. Liebman, S.A., Phillips, C., Fitzgerald, W., Wright, J., Cohen, R., Levy,

E.J., Higgins, J.H. , In *Advanced Materials: Expanding the Horizons*; Trabocco, R.; Lynch, T., Eds.; SAMPE: Covina, CA, 1992, Vol. 25; pp. 173-181. Presented at 25th Internat. SAMPE Tech. Conf., Phila. PA, Oct. 1993.

17.   Medlin, S., Morris, J.B., Fifer, R.A. *Pittcon '94,* Chicago, IL, Mar. 1994, paper no.847 "FT-Raman Spectral Interpretation using a Neural Network".

18.   Liebman, S., Ahlstrom, D., Hoke, A., *Chromatographia* **1978,** *11, 427.*

RECEIVED August 11, 1994

Chapter 3

Determination of Polymer Molecular Weight by Flow Injection Analysis and Refractive Index Gradient Detection

Separation of Simple Polymer Mixtures by Classical Least-Squares Analysis

V. Murugaiah, L. R. Lima III, and R. E. Synovec[1]

Department of Chemistry, University of Washington, Seattle, WA 98195

A method based upon flow injection analysis (FIA) is presented in which refractive index gradient (RIG) signals are used to generate temporal resolution of polymer components in a mixture without chromatographic separation. Resolution is achieved through the examination of temporal shifts in the derivative shaped signal, and in the asymmetry of that signal, which is a result of both the FIA design and RIG detection. A classical least squares (CLS) method of multicomponent analysis is used to resolve the mixtures. The RIG signal is determined to be a function of the linear combination of the weight-average concentration of the components detected. The composition of binary and ternary mixtures was resolved with better than 5% uncertainty for one trial using the CLS method. A 40%: 60% relative concentration mixture , by mass, of poly(ethylene glycol) (PEG) 1470: PEG 7100 was determined to have a relative concentration of 38.7%: 62.3% ± 1.6% at a total concentration of 150 ppm PEG by this method. Polymer component molecular weight as determined from RIG signal asymmetry is in close agreement to the molecular weight derived from an examination of an exponentially modified Gaussian peak model. A mixture of 30%: 70% PEG 1470: PEG 7100 and a total concentration of 150 ppm PEG with a weight average molecular weight of 5411 g/mol was determined to have a molecular weight of 5438 ± 106 g/mol by the asymmetry ratio method and 5443 ± 318 g/mol by CLS.

Separation, identification, and quantitation of analytes in a mixture are an ongoing challenge to analytical chemists. Chromatographic separation in different modes provides a solution to this problem (1). In particular, high performance liquid chromatography (HPLC) is a suitable separation technique for a variety of thermally labile and higher molecular weight natural and synthetic polymers (2-4). The sequence, position, and distribution of separated components contain much information about the mixture. However, HPLC experiences the disadvantages associated with the operation of pumps at higher pressures, overlapping peaks that require additional effort to achieve the required resolution, and is plagued by the general elution problem (5). Further, for high molecular weight samples, it is difficult to control the separation by gradient elution. A quantitative and rapid method that operates at low pressures is needed for the analysis of mixtures of polymers in the process environment. This new method will

[1]Corresponding author

0097–6156/94/0581–0025$08.00/0

undoubtedly involve the analysis of unresolved peaks.

Quantitation of unresolved peaks by traditional methods is time consuming and may be limited in information content. Many publications dealing with the problem of unresolved peaks can be found in the literature (6-15). Several approaches have been used to fit real chromatographic peaks to well-defined mathematical functions and these approaches find theoretical justification (8-10). Rapidly developing chemometric techniques find wide application to resolve overlapping peaks in chromatography (11-13). Recently, a novel technique for data analysis of overlapping peaks in chromatography was developed (14-15). Classical least squares (CLS) is one of the first order calibration techniques suitable for analyzing a vector of data from chromatography or flow injection analysis (FIA) (16). CLS has several advantages over traditional calibration techniques (11). First, since multiple measurements are used in calibration, a signal averaging benefit is achieved. Second, interferents or outlier samples can be detected. However, a CLS model requires prior knowledge of the pure responses of all the components present in a mixture under study.

Modeling of an analytical signal by a mathematical function enables one to obtain parameters, such as statistical moments, to characterize the peak shape of the signal. For example, a Gaussian peak model is used to derive many frequently used fundamental equations in chromatography (17), but the Gaussian function rarely provides an accurate model for real chromatographic peaks. Convectional effects introduced by the flow cell can cause asymmetry in chromatographic peaks even if a Gaussian profile has been established prior to detection (18). One model that is well accepted to represent a real chromatographic peak is an exponentially modified Gaussian (EMG) function. The FIA profile is characterized primarily by a dispersion coefficient (19) which offers information about the manifold, but direct information about the peak parameters such as second moment or variance can not be readily obtained. Recently, the EMG was used to describe FIA profiles to obtain the second moment and characterize the peak (20-22).

The molecular weight distribution of a polymer, which is often characterized by continuous monitoring of the number-average molecular weight (M_n) or the weight-average molecular weight (M_w), is essential in the characterization of the polymer (23). While many techniques are readily available and well suited for the off-line evaluation of polymer molecular weight, we are developing a technique that shows promise as a robust and rapid on-line molecular weight analyzer (MWA). The MWA should complement currently used techniques such as end group analysis by titration, light scattering, and size exclusion chromatography (SEC). Recently, we described how hydrodynamically generated concentration gradients under controlled conditions can be utilized for the determination of the molecular weight of water soluble polymers (24). In a previous study, the molecular weight of polyethylene glycols (PEGs) was determined by injecting a plug of analyte in a flow through tube and observing the radial concentration gradient detected by refractive index gradient (RIG) detection (25). Qualitative description of the mechanism for RIG detection was made by Betteridge and co-workers and utilized for detection in FIA (26). Pawliszyn and co-workers provided further description of RIG detectors for HPLC, capillary zone electrophoresis (CZE), and for electroanalytical techniques (27-31). Hancock, Renn, and Synovec demonstrated that a two order of magnitude increase in sensitivity can be gained by probing the radial concentration gradient in comparison to probing the analogous axial gradient, and have shown the utility of RIG detection in applications involving thermal gradient liquid chromatography (TGLC) (32-34). Lima and Synovec presented a model which describes the mechanism for RIG detection in relation to the dispersion of the concentration profile due to diffusional and convectional effects within the flow cell (18). The RIG detector is gaining popularity in HPLC and FIA because of good detection limits and compatibility with mobile phase and thermal gradient elution (24-37). In a previous study, the peak width and asymmetry of the RIG signal were examined and an empirical parameter, the asymmetry ratio (AR), was defined and correlated to the molecular weight of PEGs (25).

Based upon reaction chemistries, time-based selectivity in FIA was demonstrated by Whitman and co-workers by injecting a large volume of sample into a single-line FIA manifold (38). In subsequent work, comparisons of various multivariate calibration methods for data analysis in multicomponent FIA with reaction chemistry dependent, time-based selectivity in zone penetration technique was reported (39). In this study, we will utilize our previous knowledge to create reaction-independent time-resolved concentration profiles under low pressure FIA conditions, and examine the possibility of obtaining time-based resolution of a mixture of PEGs. This paper describes a method of obtaining the composition of a mixture of polymers without chromatographic separation. By examining the first derivative-based signal of a FIA profile, time-based resolution of mixtures can be achieved. With sufficient time-based resolution the time slices in a RIG detected FIA profile, which are analogous to multiple-wavelengths in a spectrum, provide a means to use CLS to obtain a unique solution for the identification of the components present in the mixture. We will show that the signal of a mixture is a linear combination of the concentration weighted response of each polymer present. Since the RIG signal obtained in our study is a first derivative of a near Gaussian or skewed Gaussian, the first derivative of an EMG is expected to model the RIG signal. Depending on the value of the decay constant of the EMG function (20-22), the asymmetry of the first derivative will vary and provide a means to extract information about the concentration profile. A simulation study is reported in which the RIG signal is modeled by the first derivative of an EMG. It will be shown that when the components present in the mixture are not known, the AR of the RIG signal provides a means to accurately estimate the weight average molecular weight of the mixture. Finally, the usefulness of the AR based molecular weight calibration will be discussed.

Molecular Weight Precision per Time. Several techniques are available for measurement of molecular weight of polymers. SEC and CZE are based on the separation of analytes and are related to the technique reported in this study. In process analysis, the analysis time and molecular weight resolution are two important parameters for consideration. In contrast to the resolution defined in separation techniques such as SEC and CZE, but identical to that used in mass spectrometry (40), we define molecular weight resolution, R_S, as given by equation 1

$$R_S = \frac{M_i}{\Delta m} \tag{1}$$

where M_i is the molecular weight of interest and Δm is its uncertainty in measurement. If S is the analytical signal of a particular technique as a function of the molecular weight, then $\frac{dS}{dM}$ is the sensitivity of the method with regard to changes in M. One may define a quality parameter, or information function, as the molecular weight resolution per unit analysis time,

$$\text{information function} = (\frac{M_i}{\Delta m})(\frac{1}{t}) \tag{2}$$

where t is the analysis time for the determination of M_i. Incorporating the uncertainty in the signal by its standard deviation S_S, which is the instrument precision level relative to $\frac{dS}{dM}$, an expression for the information function can be written as

$$\text{information function} = \left(\frac{M_i}{S_S}\right)\left(\frac{dS}{dM}\right)\left(\frac{1}{t}\right) \qquad (3)$$

The information function for related techniques are calculated, from the state of the art data reported in the references cited, and given in Table I for a target M_i of 4100 g/mol.

Table I. Molecular weight resolution and information function calculated for different methods based on the data found in references cited, for a target 4100 g/mol analyte, M_i.

Method	Time of analysis, sec	Signal sensitivity, $\frac{dS}{dM}$	Standard deviation of signal, S_S (and % relative S_S)	Resolution, $R_S, \frac{M_i}{\Delta m}$	Information Function, R_S/t sec^{-1}
CZE [a]	10	1.4×10^{-3} sec mol g^{-1}	0.22 sec(2.2%)	27	2.7
SEC [b]	75	-7.0×10^{-5} mL mol g^{-1}	0.030 mL (1%)	9.7	0.60
FIA [c]	300	1.3×10^{-2} sec mol g^{-1}	2.4 sec (0.8%)	23	0.080
FIA [d]	60	7.0×10^{-4} mol g^{-1}	0.070 (2.5%)	41	0.68

[a] Peak width based calculation, in sec (41-44).

[b] Log M versus retention volume calibration based calculation (45).

[c] Calculation based on peak width at half height, in sec (Golay equation) (47).

[d] Our method based on the asymmetry ratio of the RIG signal which is unitless (24-25).

While SEC and CZE are capable of physically resolving components, and FIA is not, the point made in Table I is that at a given M_i the reliability of a molecular weight determination improves with increasing resolution per time, R_S/t (equations 1 and 2). In addition, SEC is limited by pore size for the molecular weight resolution of analytes, as well as polymer/solvent interactions that will affect the hydrodynamic radius of the polymer, while CZE experiences the artifact of non-ideal flow rate. However, FIA-based methods operate at zero back pressure, with a lower probability of clogging than most sample modulated analyzers such as SEC. Further, FIA with RIG detection provides a means for the analysis of mixtures using multivariate statistical techniques (24).

The basis of the calculations reported in Table I will be described briefly. CZE has been shown to be a high speed, high efficiency separation technique for biopolymers (41-42). CZE can be used to determine the diffusion coefficient, D, for analytes and hence their molecular weight (43). Temporal peak variance determined in CZE was converted into a length variance and the Einstein equation was then used to calculate the D (43). In CZE, the major source of uncertainty is the lack of reliability of migration times, at a constant observation distance, and routinely this was found to be on the order of 10%, while state-of-the-art is 2.2% (44). The uncertainty in retention time in CZE may be reduced by using markers of higher and lower mobility than the analyte, but with a longer analysis time as well. The uncertainty in migration time, dt, results in an uncertainty, dM, in the estimated molecular weight. So, in CZE, $\frac{dt}{dM}$ is the sensitivity $\frac{dS}{dM}$. The migration time uncertainty in CZE reduces the information function (Table I).

In SEC, a plot of elution volume versus log M is a straight line, within a limited molecular weight range. Hence, the sensitivity, $\frac{dS}{dM}$, can be evaluated from the standard calibration curve of log M versus retention volume (3,45). Uncertainties in retention volume limits the M_i prediction in SEC. The uncertainty in M_i is improved by appropriate choice of stationary phase pore size. Using FIA techniques, the peak width of the concentration profile obtained under a suitable condition is related to the diffusion coefficient of the analyte, and hence to the molecular weight (5,46). According to FIA theory presented by Vanderslice and co-workers, the role of convection and diffusion on the detected concentration profile can be predicted using knowledge of the tube radius, r, tube length, L, flow rate, F, the analyte molecular weight, and D (47). The concentration profile is predicted by calculating the reduced time (τ), a dimensionless quantity given by

$$\tau = \frac{Dt}{r^2} = \frac{\pi DL}{F} \tag{4}$$

with t being the transit time through the tube. For $\tau < 0.04$, the dispersion of the solute is dominated by convection, while for $\tau > 0.4$, diffusion dominates. Flow cells used for RIG detection possess τ values on the order of 0.001, so convection dominates the dispersion of the analyte. Fortunately, the analyte spends very little time in the flow cell, so the net affect on the concentration profile is minimal, introducing only a small degree of asymmetry to the detected peak. For $0.04 \leq \tau \leq 0.4$, convection and diffusion more equally interplay to control the shape of the concentration profile. Under the conditions of $\tau > 0.4$ a Gaussian profile is obtained and one can use the Golay equation to define diffusion and obtain the molecular weight from the peak-width measurement, but working in the $\tau > 0.4$ range will unduly increase the analysis time relative to the $0.04 \leq \tau \leq 0.4$ range, all other conditions being constant (24). Alternatively, one can decrease the radius of the flow through tube to increase τ, keeping other conditions the same.

FIA with RIG detection for $0.04 \leq \tau \leq 0.4$. Based upon the data shown in Table I and previous discussion, we will proceed with the description of FIA with RIG detection method. A typical RIG signal is given in Figure 1. A RIG signal is characterized by a leading peak height $\theta(+)$, a trailing peak height $\theta(-)$, a leading peak width $W(+)$, and a trailing peak width $W(-)$. For each RIG signal, as in Figure 1, three time points are defined: $t(+)$ is the time between injection and $\theta(+)$, $t(0)$ is the time taken to reach the zero crossing of the baseline, and $t(-)$ is the time taken to reach $\theta(-)$. In a previous study (25), the dependence of the peak widths of RIG signals to the molecular weight of PEGs was investigated. Since the RIG signal is a measure of the radial

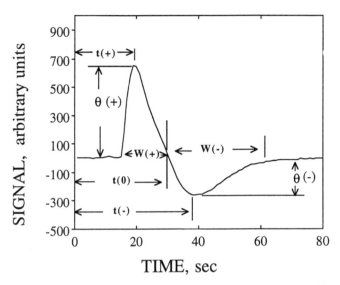

Figure 1. Refractive index gradient signal of PEG 7100. θ(+), leading signal; q(-), trailing signal; W(+), leading peak width; W(-), trailing peak width; t(+) is the time between injection and θ(+); t(0) is the time taken to reach the zero crossing of the base-line; and t(-) is the time taken to reach θ(-).

concentration gradient (18,24) it approximates the shape of the first derivative of a nearly Gaussian profile. Peak width is a function of the analyte diffusion coefficient. Thus, peak shape depends on how far the analyte diffuses in the flow stream. Polymers with large molecular weight (and hence a lower diffusion coefficient) have widely different leading and trailing peak widths that approach a common value as the analyte diffusion coefficient increases (25). A quality parameter called the asymmetry ratio, AR, was defined as

$$AR = \frac{\theta(+)}{\theta(-)} \tag{5}$$

The AR was correlated to the diffusion coefficient, and hence to the molecular weight (25). In previous studies (24-25) the relationship between the AR and weight average molecular weight, M_w, was nearly linear over the range of PEGs examined as a function of various instrumental conditions such as flow rate, tube length, and so forth, and is given by

$$AR = m M_w + y \tag{6}$$

where m and y are calibration constants. An important observation was that the AR was independent of the concentration of PEG from the detection limit of about one ppm up to about 1000 ppm (25). Infinite dilution conditions are defined as the concentration range over which the analyte diffusion is invariant (48), so the solutions examined in this discussion are considered to be at infinite dilution.

The concentration profile obtained by the diffusion and convection mechanism in FIA is well characterized by the EMG (20-21). The EMG peak is defined by the following standard convolution integral.

$$f(t) = \frac{A\sigma}{\tau\sqrt{2}} \exp[\frac{1}{2}\left(\frac{\sigma}{\tau}\right)^2 - \left(\frac{t - t_R}{\tau}\right) \int_{-\infty}^{Z} \exp[-x^2]\, dx \qquad (7)$$

where the limit of integration, Z, is defined as

$$Z = \frac{1}{\sqrt{2}}\left[\frac{(t - t_R)}{\sigma} - \frac{\sigma}{\tau}\right] \qquad (7A)$$

In equation 7, A is the peak amplitude, σ is the variance of the Gaussian profile, t_R is the center of gravity of the Gaussian, τ is the time constant of the exponential modifier and x is a dummy variable of integration (9). The asymmetry of the EMG peak depends on the ratio τ/σ. The integral term from equation 7 can be approximated by the error function (9) given by

$$\text{erf}\,\frac{1}{\sqrt{2}}\left(\frac{t_R}{\sigma} + \frac{\sigma}{\tau}\right) + \text{erf}\frac{1}{\sqrt{2}}\left[\frac{(t - t_R)}{\sigma} - \frac{\sigma}{\tau}\right] \qquad (8)$$

Behavior of the First Derivative of the EMG. The EMG functions were generated by suitably selecting τ/σ values and differentiating them to obtain the first derivative which resembled the RIG signals. Further, the differentiation process

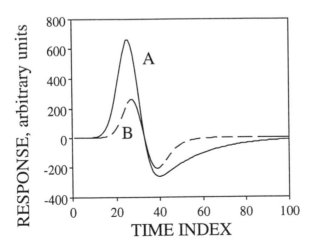

Figure 2. First derivative traces of the EMG model simulation, using equations 7 and 8. A: For $\tau/\sigma = 4$ providing an asymmetry ratio (AR) of 2.52; B: for $\tau/\sigma = 1$ providing an asymmetry ratio (AR) of 1.25 according to equation 5.

magnifies the noise in the simulated data set. A simplified least square procedure suggested by Savitzky and Golay (49) was used to smooth the first derivative of the EMG function given by equation 7.

Figure 2 illustrates two simulated signals A and B with an AR, calculated as in equation 5, of 1.25 and 2.52 corresponding to τ/σ values of 1 and 4 respectively. These two signals would result from different polymers, in this example corresponding to PEG 1470 and PEG 7100, under the experimental conditions to be described later. The progress of elution of the overall concentration profile may be characterized by the center of gravity of the profile or mean arrival time defined as

$$\text{mean arrival time} \quad = \quad \frac{\displaystyle\sum_{t=1}^{N} |\theta(t)| \, t}{\displaystyle\sum_{t=1}^{N} |\theta(t)|} \tag{9}$$

where $|\theta(t)|$ is the absolute value of the detector response at the tth time index, t is the time index measured from the point of injection, and N is the total time index. Though the center of gravity of the profiles remain essentially the same, the peak shapes of the two profiles in Figure 2 are significantly different from one another because of the differences in time, namely t(+) and t(-) and peak widths W(+) and W(-) as defined in Figure 1. The data in Figure 2 can be used to illustrate the effective time-based resolution of these two signals. If the two signals represent the RIG signals of two PEGs present together in a sample, a resolution, analogous to the chromatographic separation of two components, can defined. With reference to Figure 1 and Figure 2, the effective time based-resolution, R_t, of the leading peak is given by

$$R_t = 2 \frac{(t_A(+) - t_B(+))}{(W_2 + W_1)} \tag{10}$$

where $t_A(+)$ and $t_B(+)$ are the times taken for the leading profiles to peak-out on separate injections of two analytes A and B, and W_2 and W_1 are the respective peak widths. Since the leading peak is more sensitive than the trailing peak, the leading peak width is used in equation 10. For the example illustrated in Figure 2, R_t is 0.16 for the leading edge peak. Although the effective time-based resolution between two RIG signals is not as large as one might expect in a chromatographic separation, it is large enough to provide a means of resolving RIG signals of mixtures containing a few components.

It is important to note that the properties of the first derivative of a suitable EMG function were qualitatively compared with the properties of the RIG signal and no attempt was made to fit the RIG signal data exactly to the function, although the agreement is significant. The asymmetry introduced to the Gaussian profile by the different exponential decay constants results in profiles which differ in the relative positions of the maxima, minima, and the zero-crossing point of the first derivative of the EMG. The value of t(0) was essentially constant for all simulated profiles, according to equation 9, which is consistent with FIA experiments (19). Figure 3 shows the times t(+), t(0), and t(-) as defined earlier for the first derivative of EMG after centering to the zero-crossing point, specifically setting the relative position of t(0) to zero, for different values of τ/σ ranging from 1 to 3.5. The invariant nature of t(0) indicates that the center of gravity of these profiles, given by equation 7, remain the same as long as the underlying concentration profile has the same retention, t_R, as will be the case in the unretained FIA experiment. Meanwhile, t(+) and t(-) changed

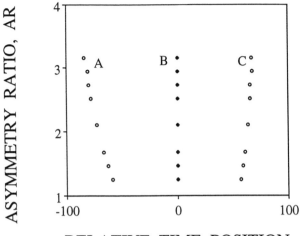

RELATIVE TIME POSITION

Figure 3. Asymmetry ratio versus the relative time position of the simulated first derivative of EMG for selected values of τ/σ ranging from 1 to 3.5, using equations 1, 3, and 4. A = t(+); B = t(0); C = t(-); with the time index t(0) adjusted from its original constant value to zero.

substantially, thus affording some "resolution" through equation 10. With increasing asymmetry, the first derivative of the EMG function gives a sharp rise which is analogous to the RIG signal of a high molecular weight polymer.

Concentration gradient profiles for synthetic mixtures were generated and the AR calculated to predict the weight average molecular weight of the mixture. Let the profiles A and B in Figure 2 represent the signals of polymers with molecular weight M_A and M_B, respectively. A series of composite profiles, pA + (100-p)B, where p is the percentage of A in the mixture, was simulated with p ranging from 0 to 100 by multiples of 5. The AR of the mixture profile was calculated and the molecular weight predicted by equation 6, with m = 2.2×10^{-4} and y = 1.01, for M_A = 1470 and M_B = 7100, and the result referred to as $M_{W,AR}$. These calibration constants are consistent with our previous work, as well as the experimental data in Figure 6B to be discussed shortly. In addition, the corresponding weight average molecular weight of the mixture was calculated as $[pM_A + (100-p)M_B]/100$ for all the simulated mixtures and referred to as $M_{W,fit}$. Figure 4A is a plot of $M_{W,AR}$ versus $M_{W,fit}$ with a reference line of slope one, indicating the closeness of the AR based molecular weight with the weight average molecular weight, $M_{W,fit}$. Figure 4B shows the percentage of deviation of $M_{W,AR}$ with respect to $M_{W,fit}$. The non-linear prediction has the greatest deviation in the middle of the range of molecular weights examined. However, the error of prediction of the molecular weight of the mixture by the AR method is less than 5%.

Alternatively, if the pure responses of the components present in a mixture are known, CLS may be used to determine the contribution of individual components. As in our previous reports (24-25), the RIG signal used in this study measures the radial concentration gradient. For dilute solutions of PEGs, the RIG signal at any time point i is proportional to the radial concentration gradient which in turn is a function of the axial concentration profile (24). In the absence of any interaction between PEGs, in the so called "infinite dilution" region (25,34), the RIG signal of a mixture should be a linear combination of the component signals, given by the model

$$R_i = \sum_{j=1}^{S} f_j \theta_{ij} + \varepsilon_i \qquad (11)$$

where R_i is the response of the mixture at a time slice i, f_j is the weight fraction of the jth component analyte in the mixture and θ_{ij} is the pure response of the jth analyte at time i, and ε_i is a random error term at the ith time slice and S is the total number of the

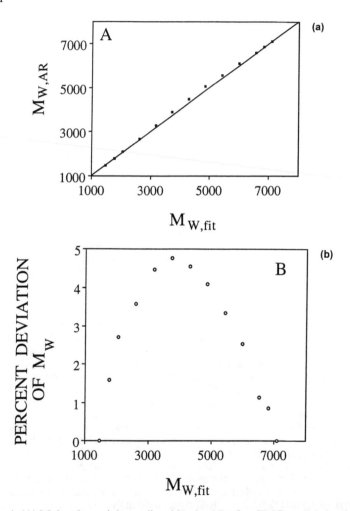

Figure 4. (A) Molecular weight predicted by the AR of an EMG model simulated signal, $M_{W,AR}$, versus the weight-averaged molecular weight, $M_{W,fit}$, of mixtures containing PEG 1470 and PEG 7100, superimposed on a line of slope 1 to indicate the relative deviation. (B) Percentage deviation of the AR based molecular weight with respect to the weight-averaged molecular weight predicted by CLS for EMG model simulated signals.

components present in the mixture. The weight fraction of the component present in the mixture is considered to be the appropriate concentration scale for the RIG detector response.

In matrix notation, the CLS model that describes the given experimental design is

$$\mathbf{R} = \theta \, \mathbf{f} + \epsilon \tag{12}$$

where $\mathbf{R}$ is a Nx1 response vector for a mixture containing S components, N is the total time slices used in the calculation, θ is the NxS matrix of pure PEG signals, $\mathbf{f}$ is a vector of the weight fraction of the components in the mixture and ϵ is the error vector. R and θ are experimentally determined. The least square estimate of $\mathbf{f}$ is given by

$$\hat{\mathbf{f}} = (\theta^T \theta)^{-1} \theta^T \mathbf{R} \tag{13}$$

where superscript T means the transpose of the matrix θ and $\hat{\mathbf{f}}$ is an estimate of the fraction of the components present in the mixture. Further, the least square estimate of $\mathbf{f}$ was obtained with the restriction that only positive values of $\hat{\mathbf{f}}$ were allowed using the non-negative least square algorithm. Agreement of the estimated fractions of the components present in a mixture with experimental values offers a means to quantitatively evaluate the unknown in the mixture provided all the analytes in the mixture are known. The students'-t test may be used to test the agreement. In addition, the goodness of fit can be evaluated by calculating the residual sum of squares (RSS) defined as

$$RSS = \sum_{i=1}^{N} (\mathbf{R} - \theta \hat{\mathbf{f}})^2 \tag{14}$$

Further, the accuracy of the estimated fraction as compared to the true fraction of the component present in the mixture is also important. The error in the estimation of the weight fraction, f_j, is evaluated by the relative root mean square error (RRMSE), or the coefficient of variation,

$$RRMSE = \frac{100}{\bar{f}} \sqrt{\frac{\sum_{j=1}^{r} (\hat{f}_j - f_j)^2}{d.f.}} \tag{15}$$

In equation 15, $\bar{f}$ is the mean of the weight fraction of a component in the mixture, $\hat{f}_j$ is the predicted weight fraction of the component from the jth replicate, r is the number of replicates, and d.f. denotes the degree of freedom in the calculation of error. The relative error is given in percent.

Experimental. The apparatus used in this work is similar to that of our previous work (23) and is not shown for brevity. The 633 nm, 5 mW cw output from a He-Ne laser (model 105-1, Uniphase, San Jose, CA) was focused by a 20X microscope objective (Newport Corporation, Fountain Valley, CA) on to a single-mode optical fiber (4 μm core, 125 μm clad, 250 μm jacket) (Newport Corporation, Fountain Valley, CA) which is optimized for wavelength of 633 nm. The optical fiber was mounted on a precision fiber coupler with a fiber chuck (Newport Corporation, Fountain Valley, CA). A collimated probe beam was produced by interfacing the optical fiber output to a GRIN lens (SELFOC lens, NSG America, Sommerset, NJ) that was quarter pitch at 633 nm. The narrow, collimated probe beam, 200 μm in diameter at the flow cell entrance, was passed through a Z-configuration flow cell, made-in-house, with a 10 cm distance separating the flow cell and the GRIN lens. The flow cell had a 1.0 cm length and 800 mm diameter with 1/16 in o.d. x 0.007 in i.d. PEEK tubing (Upchurch Scientific Inc., Oak Harbor, WA), and possessed a t value on the order of 0.001. The flow cell was mounted on a high precision X-Y-Z translational stage (Newport Corporation, Fountain Valley, CA) which allowed radial adjustment of the position of the beam (18,24). After the collimated laser beam passed through the flow cell, it was sent onto a bi-cell position sensitive detector (PSD) (Hamamatsu City, Japan). The optical and mathematical configuration of the PSD device was discussed in detail elsewhere (50). The ratio output of the PSD is independent of the incident beam intensity. The beam deflection due to the RIG was proportional to the PSD ratio output.

Deionized water served as the carrier and was delivered by a syringe pump (ISCO, model 2600, Lincoln, NE) through an automated injection valve fitted with a 2.5 μL injection loop and a two-position electric actuator (Valco Instruments Co. Inc., Houston, TX). Since there can be segregation of PEG, due to polydispersity, semi-solid PEG was melted in a hot water bath and flake sample was crushed and well mixed to get a representative sample for weighing. Approximately 150 ppm aqueous solutions of all PEGs listed in Table II were prepared from stock solutions of 3,000 ppm.

TABLE II.   Polyethylene glycol standards examined.

$M_W{}^a$ (g mol^{-1})	Label[b]
1470	PEG 1470
4100	PEG 4100
7100	PEG 7100
8650	PEG 8650

[a] Weight average molecular weight, M_w quoted by the manufacturer, Polymer Sciences Laboratories Inc., Amherst, MA. The polydispersity quoted by the manufacturer is quite low, ranging from 1.03 to 1.06 for these polymers.

[b] Label used for discussion in text.

For studies of the multicomponent resolution of a mixture of PEGs, a 75 cm length of 0.007 in. i.d. PEEK tubing was connected between the injection valve and the flow cell. This tubing was longer than used in previous reports (24), and was chosen to allow examination of smaller D via equation 4 while keeping the τ range constant, and to minimize the dispersion introduced by the flow cell. Dispersion within the flow cell used is dominated by convection (18) (τ on the order of 0.001) and introduces asymmetry to the detected signal. However, the flow cell has a small length, so the convectional effects can be minimized by using a longer diffusion dominated dispersion tube prior to the flow cell.

The laser probe beam was set at the optimum position, 58% of the distance from the flow cell center to the wall on the vertical axis, by carefully positioning the flow cell X-Y-Z translational stage vertically relative to the probe beam (24). A flow rate of 50 μL/min was used with five replicate injections made for each PEG standard and mixture.

The RIG signals were recorded on a chart recorder and collected via a laboratory interface board (IBM Data Acquisition and Control Adapter, IBM, Boca, Raton, FL) with a personal computer (Leading Edge, model D, Canton, MA).

Data Processing and Simulations. Data were collected at 20 Hz for 2.5 minutes, resulting in 3000 time channels or slices per injection. The data from each sample was baseline corrected and truncated to retain only 700 time slices of FIA profile. This was accomplished by discarding the first 300 data points and the last 1300 data points. Error function resident in the MATLAB program (The MathWorks, Inc., South Natick, MA) was used to evaluate the EMG function according to equation 7 with the approximation for integral term as given by equation 8. τ/σ values were selected in order to get the first derivative of EMG peaks to resemble as close to the RIG signals. By direct comparison of the simulated signals in Figure 2 with the AR of the PEGs, it was found that PEG 1470 and PEG 7100 are represented by the first derivative of the EMG with τ/σ of 1 and 2.52 respectively.

Results and Discussion. One of the primary considerations of this study is to show that single-line, single-injection FIA can be used to create non-chromatographic, yet time-resolved profiles for multicomponent analysis.

Experimental conditions such as the flow through tube length and flow rates can be manipulated to obtain RIG signals of different asymmetry to obtain molecular weight information on a series of PEGs (24). A manifestation of the diffusion of PEGs in a flow through tube and flow cell results in temporally resolved peaks under suitable experimental conditions (25). In the study to resolve a mixture of PEGs, the experimental conditions were selected to gain wide differences in t(+) between PEG 1450 and PEG 8650. Figure 5 shows the RIG signals of PEG 1470 and PEG 7100 which is comparable to the data in Figure 2 obtained by simulation using the EMG model. The profiles of PEG 1470 and PEG 7100 show distinct peak shape differences in the leading and trailing portions but maintain the same center of gravity of the profiles. The mean arrival time for all the PEGs, calculated according to equation 9, are essentially the same but the zero crossing time is slightly smaller for each PEG,

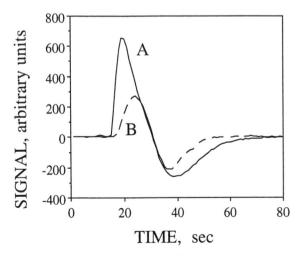

Figure 5. Experimentally measured RIG signals used to demonstrate the time-based resolution, R_t. R_t was found to be 0.27 for the leading peak, according to equation 10. (A) Solid line, PEG 7100; (B) dashed line, PEG 1470.

indicating that the concentration profile is skewed. Note that the calculation of mean arrival time is sensitive to the truncation of the data set.

Though the mean arrival time is the same for different polymers, an effective resolution based on the shape of the leading peaks can be calculated using equation 10. Under the experimental conditions reported, a time-based resolution of 0.27 was calculated between PEG 1470 and PEG 7100 (Figure 5). Though a resolution of 0.27 seems small from conventional HPLC separation standards (1), various chemometric techniques can be used to analyze poorly resolved peaks to about 0.10 resolution (12-15).

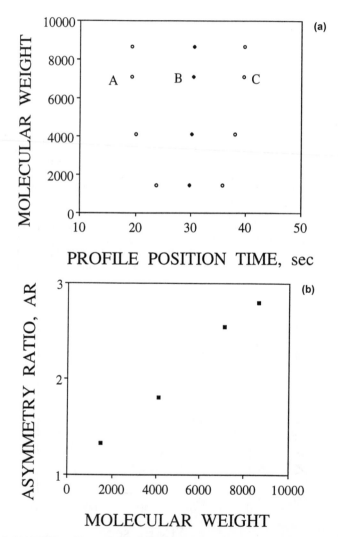

Figure 6. (A) RIG profile position times for PEGs given in Table II, under the experimental conditions described in the text. A = t(+); B = t(0); C = t(-); see Figure 1 for definition of symbols. (B) Asymmetry ratio as a function of molecular weight for the PEGs.

RIG signal sensitivity is proportional to $\dfrac{1}{D/u + C}$, where C is the system dependent, analyte independent, contribution due to convective dispersion of the concentration profile and u is the average linear flow velocity of the solvent through the flow cell (18). At the linear flow velocities examined both the $\dfrac{D}{u}$ and C terms contribute significantly, so one would expect a more sensitive peak for PEG 7100 for both leading and trailing peak portions, as is observed in Figure 5, due to a smaller value of D, than for PEG 1470 even though the concentrations of both samples are nearly equivalent. The different times, t(+), t(0), t(-), as defined earlier in Figure 1, for all PEG standards studied are given in Figure 6A. The leading sharp peak of the larger polymer with small diffusion coefficient arrives at a shorter time than for the smaller polymer. The timing of the RIG profiles given by t(+), t(-), and t(0) parallel the relative time index of the simulated profiles obtained by the first derivative of the EMG as given in Figure 3. Since the RIG detector measures the bulk property of the concentration gradient directly, the response at any time point is a function of the diffusion coefficient of the individual analyte present, and the overall response should be given by the weighted sum of the responses of the individual components present as given by equation 11. Because of the presence of time-based resolution discussed previously, there should be a time-based selectivity comparable to the wavelength selectivity in an absorbance detection, which enables one to resolve the component peaks in a mixture. If there is no time-based selectivity, the response of one PEG will almost be a scalar multiple of the response of another PEG and the pseudo inverse matrix, $(\theta^T \theta)^{-1}$, in equation 13 will be "ill

Table III. Classical least-squares (CLS) results for fitting mixture response versus pure components response using the responses of all the components present as well as with an additional component.

Calculation number	Standards used in CLS	Relative amount present in sample (%)	Amount predicted (%) ± s.d.	RRMSE [a] (%)	RSS[b]
1	PEG 1470	40.0	38.7 ± 2.0	5.1	5.4×10^4
	PEG 7100	60.0	62.3 ± 1.1	1.7	
2	PEG 1470	40.0	38.3 ± 1.5	3.9	
	PEG 4100	0	1.7± 3.5	205.8	8.6×10^4
	PEG 7100	60.0	62.6 ± 3.5	5.6	
3	PEG 1470	33.3	34.0 ± 5.7	16.7	
	PEG 4100	33.3	33.5 ± 3.0	8.9	7.9×10^4
	PEG 7100	33.3	33.5 ± 2.3	7.1	
4	PEG 1470	33.3	40.0 ± 11.0	27.5	
	PEG 4100	33.3	29.0 ± 17.0	58.6	6.0×10^4
	PEG 7100	33.3	19.0 ± 20.0	105.3	
	PEG 8650	0	16.0 ± 15.0	93.7	

[a] RRMSE = Relative root mean square error.

[b] RSS = Residual sum of squares.

s.d.   = Standard deviation.

conditioned" because of the colinearity problem resulting in poor resolution (11). Figure 6B shows the AR as a function of the molecular weight. It is important to note that PEG 7100 and PEG 8650 have negligible differences in t(+), as shown in Figure 6A, but show considerable differences in AR. Therefore, when the CLS fails to resolve a mixture of PEG 7100 and PEG 8650 because of the colinearity problem, the AR can be used to predict the molecular weight of the mixture, as will be described.

The condition for using CLS regression for calibration is that the concentrations of all species giving a non-zero response must be included in the calibration step. In the study to use the CLS method, various mixtures were studied as follows: three binary mixtures containing PEG 1470 and PEG 7100 and a ternary mixture containing equal amounts of PEG 1470, PEG 4100 and PEG 7100. Results are summarized in Table III. Figure 7 shows the RIG profiles of a 4:6 mixture of both PEG 1470 and PEG 7100, whereas the RIG profiles for PEG 1470 and PEG 7100 are shown in Figure 5. For the binary mixture, the non-negative least square fit to the observed mixture response was performed using the pure responses of PEG standards, using equation 13. The residuals from the fit (Figure 7) were examined. The predicted weight percentage was statistically in good agreement with the amount present, as shown in Table III, calculation number one. In order to study the uniqueness of this fit, the responses of the binary mixture containing PEG 1470 and PEG 7100 were fitted to a matrix of pure responses of PEG 1470, PEG 4100, PEG 7100, and it yielded the expected amount of PEG 1470 and PEG 7100 and almost zero contribution from PEG 4100, as shown in Table III, calculation number two. Hence, when there is time-based resolution, the CLS fit provides a unique solution providing almost negligible contribution from components not present in the mixture.

The ternary mixture response was fitted with the knowledge of the pure responses of the components present. Table III provides the summary of the results obtained. The predicted amount is in good agreement with the amount present as shown

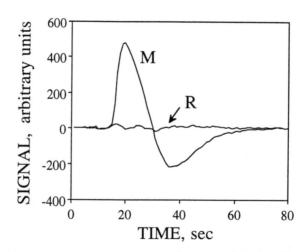

Figure 7. M: RIG response of a 40:60 mixture of PEG 1470 and PEG 7100, with individual PEG response shown in Figure 5. R: Residuals from CLS fit of data in Figure 5 to M.

in calculation number three. However, when PEG 8650 is included in the training set for prediction, the error of prediction increases and the algorithm does not provide a reasonable unique solution as shown in calculation number four. This behavior is anticipated from the fact that although there is an AR difference between PEG 7100 and PEG 8650, there is no time resolution between these two profiles. With respect to the CLS model, PEG 7100 and PEG 8650 profiles are nearly a multiple of one another resulting in a poor prediction if both are present in the CLS training set.

Conclusions. For a mixture of polymers, the average molecular weight reported depends on how it is calculated or measured (23). When there is time-based resolution, the RIG response of a mixture can be resolved into its component weight fractions. Furthermore, it is possible to use the AR of the mixture RIG profile to accurately estimate the weight average molecular weight of PEG mixtures when resolution is not possible with CLS as is clear from the results in Table IV for the three PEG 1470 and PEG 7100 mixtures.

Table IV. Molecular weight calculated from the classical least-squares (CLS) fit coefficients of mixture response according to equation 13 and molecular weight evaluated from the calibration of asymmetry ratio of mixture profiles according to equation 6.

Composition of mixture PEG 1470 and PEG 7100	$M_{w,abs}$	$M_{w,AR} \pm$ s.d.	$M_{w,fit} \pm$ s.d.
30:70	5411	5438 ± 106	5443 ± 318
40:60	4848	5134 ± 180	5044 ± 145
50:50	4285	4605 ± 230	4476 ± 220

$M_{w,abs}$ = Calculated from known composition, weight-average.
$M_{w,AR}$ = Molecular weight based on asymmetry ratio of the RIG signals.
$M_{w,fit}$ = Molecular weight based on CLS results.

Thus, for the PEG samples studied, when it may not be possible to resolve a mixture to obtain its composition, the AR of the RIG signal provides a means of accurately estimating the weight average molecular weight to within a 5% or less error, typically, as illustrated in Fig 4B. Note that, in general, this FIA method for molecular weight determination is highly dependent upon the detector selectivity with respect to the specific polymer molecular weight and chemical functionality. Correlation of detected peak shape to polymer mixture molecular weight average must be carefully evaluated, but the potential benefit of developing a FIA method may be advantageous as a quality control tool.

Acknowledgment. We gratefully acknowledge the support of this work by Union Carbide Corporation and helpful discussions with Jerry R. Hale. We also thank B.R. Kowalski and co-workers of the Center for Process Analytical Chemistry for their helpful discussions and comments.

Literature Cited:
1.   Karger, B. L.; Snyder, L. R.; Horvath, Cs. *Introduction to Separation Science*, Wiley: New York, 1973.
2.   Snyder, L. R.; Kirkland, J. J. *Introduction to Modern Liquid Chromatography*, 2nd ed.; Wiley: New York, 1979.
3.   Yau, W. W.; Kirkland, J. J.; Bly, D. D. *Modern Size-Exclusion Liquid*

Chromatography: Practice of Gel Permeation and Gel Filtration
Chromatography, Wiley: New York, 1979.

4. Belenkii, B. G.; Vilenchik, L. Z. Modern Liquid Chromatography of
Macromolecules, Elsevier: Amsterdam, Vol. 45, 1983.

5. Giddings, J. C.; Davis, J. M.; Schure, M. R. Ultrahigh Resolution
Chromatography, ACS Symposium Series: Washington D.C. 1984, 250 9.

6. Anderson, A. H.; Gibbs,T. C.; Littlewood, A. B. Anal. Chem. 1970, 42
434.

7. Anderson, A. H.; Gibbs,T. C.; Littlewood, A. B. J. Chromatogr. Sci. 1970,
8, 640.

8. Grushka, E. Anal. Chem. 1972, 424, 1733.

9. Jeansonne, M. S. J.; Foley, J. P. J. Chromatogr. Sci. 1991, 29, 258.

10. Sternberg, J. C.; in Giddings, J. C.; Keller, R.A. Eds., Advances in
Chromatography, Vol. 2, Marcel Dekker: New York, 1966.

11. Sharf, M. M.; Illman, D. L.; Kowalski, B. R. Chemometrics, Wiley-
Interscience: New York, 1986.

12. Ramos, L. S.; Sanchez, E.; Kowalski, B. R. J. Chromatogr. 1987, 385, 165.

13. Lindberg, W.; Oehman,; W.; Wold, S. Anal. Chem. 1986, 58, 299.

14. Synovec, R. E.; Johnson. E. L.; Bahowick. T. J.; Sulya, A. E. Anal. Chem.
1990, 62, 1597.

15. Bahowick, T. J.; Synovec, R. E. Anal. Chem. 1992, 64, 489.

16. Beebe, K. R.; Kowalski, B. R. Anal. Chem. 1987, 59, 1007A.

17. Grushka, E.; Myers, M. N.; Schettler, P. D.; Giddings, J. C. Anal. Chem.
1969, 41, 889.

18. Lima, III, L. R.; Synovec, R. E. Anal. Chem. 1993, 65, 128.

19. Ruzicka, J.; Hansen, E. H. Flow Injection Analysis, Wiley:
New York, 1988.

20. Brooks, S. H. ; Leff, D. V.; Hernandez Torres, M. A.; Dorsey, J. G. Anal.
Chem. 1988, 60, 2737.

21. Brooks, S. H.; Dorsey, J. G. Anal. Chim. Acta 1990, 229, 35.

22. Valcarcel, M.; Luque de Castro, M. D. Flow Injection Analysis: Principles and
Applications, Ellis Horwood: Chichester, U.K., 1987.

23. Cambell, D.; White, J. R. in Polymer Characterization, Chapman and Hall:
New York, 1989.

24. Murugaiah, V.; Synovec, R. E. Anal. Chem. 1992, 64, 2130.

25. Murugaiah, V.; Synovec, R. E. Anal. Chim. Acta 1991, 246, 241.

26. Betteridge, D.; Dagless, E.L.; Fields, B.; Graves, N.F. Analyst, 1978, 103,
1230.

27. Pawliszyn, J. Anal. Chem. 1986, 58, 243.

28. Pawliszyn, J. Anal. Chem. 1986, 58, 3207.

29. Pawliszyn, J. Spectrochimica Acta Rev. 1990, 13 , 311.

30. Wu, J and Pawliszyn, J. Anal. Chem. 1992, 64, 2934.

31. Pawliszyn, J. Anal. Chem. 1992, 64 , 1552.

32. Hancock, D. O.; Synovec, R. E. Anal. Chem. 1988, 60, 1915.

33. Hancock, D. O.; Renn, C. N.; Synovec, R. E. Anal. Chem., 1990, 62, 2441.

34. Renn, C. N.; Synovec, R. E. J. Chromatogr. 1991, 536, 289.

35. Murugaiah, V.; Synovec, R. E. SOQUE Lasers '90 Conference Proceedings:
Lasers in Chemistry, 1991, 13, 763.

36. Hancock, D. O.; Synovec, R. E. Anal. Chem. 1988, 60, 2812.

37. Hancock, D. O.; Synovec, R. E. J. Chromatogr. 1989, 464, 83.

38. Whitman, D. A.; Christian, G. D.; Ruzicka, J.Anal. Chim. Acta, 1988, 214,
197.

39. Whitman, D. A.; Seasholtz, M. B.; Christian, G. D.; Ruzicka, J.;
Kowalski, B.R.Anal. Chem., 1991, 653 , 775.

40.   White, F. A.; Wood, G. M. *Mass Spectrometry: Application in Science and Engineering*, John Wiley & Sons, 1986.
41.   Jorgenson J. W.; Lukacs, K. D. *Anal. Chem.*, **1981**, *53*, 1298.
42.   Monnig, C. A.; Jorgenson J. W. *Anal. Chem.*, **1991**, *63*, 802.
43.   Walbroehl, Y.; Jorgenson J. W. *J. Microcolumn Sep.* **1989**, *1*, 41.
44.   Chang, H.; Yeung, E. S. *Anal. Chem.* **1993**, *65*, 650.
45.   Renn, C. N.; Synovec, R. E. *Anal. Chem.* **1988**, *60*, 200.
46.   Boyle, W. A.; Buchholz, R. F.; Neal, J.A.; McCarthy, J.L; *J. Appl. Poly. Sci.*, **1991**, *42*, 1969.
47.   Vanderslice, J. T.; Rosenfeld, A.; Beecher, G. R. *Anal. Chim. Acta* **1986**, *179*, 119.
48.   *Polymer Handbook*, 3rd ed., Wiley: New York, 1989 p. Vol. VII /62.
49.   Savitzky, A.; Golay, M. J. E. *Anal. Chem.* **1964**, *36*, 1627.
50.   Renn, C. N.; Synovec, R. E. *Anal. Chem.* **1988**, *60*, 1188.

RECEIVED September 6, 1994

Chapter 4

Structure and Orientation of Surfactant Molecules at the Air–Water Interface

L. J. Fina[1], J. E. Valentini[2], and Y. S. Tung[1]

[1]Department of Mechanics and Materials Science,
College of Engineering, Rutgers University, Piscataway, NJ 08855
[2]DuPont, Imaging Systems Department, Brevard, NC 28712

Fourier transform reflection-absorption infrared (RA-IR) spectroscopy is used to probe the structure and properties of sodium dodecyl sulfonate ($C_{12}S$) monolayers that are self-assembled from dilute solution at an air-water interface. Recent optical models for the interpretation of signal intensity measurements are briefly reviewed. The methylene stretching peaks of $C_{12}S$ monolayers in the RA-IR spectra are used to determine the chain orientation, the surface concentration and the conformational state of the alkyl chains. Comparisons are drawn between monolayers and $C_{12}S$ crystals. A phase transition is found as the concentration of $C_{12}S$ in the subphase below the monolayer is reduced. The effect of salt on the monolayers is presented. The infrared data is interpreted in terms of the surface tension behavior.

A renewed interest in the structure and properties of amphiphilic molecules which reside at the air-water interface has occurred in the last decade due to the fact that the monolayers comprise an idealized two dimensional system which can be probed in terms of structure, composition and phase transitions, and due to the fact that the monolayers are used to form highly ordered coatings in Langmuir-Blodgett applications. Monolayers form at an air-water interface upon dissolution of amphiphiles in a water immiscible solvent, spreading of the solution on a water surface, and lastly, solvent removal. Such monolayers are insoluble in the aqueous phase. Alternately, monolayers which are soluble in the aqueous phase are formed by adsorption of amphiphiles from the bulk aqueous solution to the air-water interface. A primary difference between insoluble and soluble monolayers

0097–6156/94/0581–0044$08.00/0

is a molecular exchange that only occurs between the monolayer and the subphase in soluble systems. This leads to an adsorption equilibrium where the bulk phase concentration of the amphiphile determines the surface tension.

Most of the techniques applied to *in situ* monolayers at the air-water interface have been used to probe the structure and properties of insoluble systems. Surface pressure-area isotherms are often measured for an initial view of phase transition behavior. A graphical illustration of the coexistence of phases in phospholipids has been provided with fluorescence microscopy. Micrographs show circular phase domains in the flat part of surface pressure-area isotherms (*1-3*). Further, methods with high spacial and temporal resolution such as surface potential measurements and ellipsometry show large signal fluctuations as the islands traverse through the probed area (*4,5*). Signal fluctuations occur in molecular area regimes of phase coexistence, and disappear in regimes of pure phase. Much smaller island structures have been measured with synchrotron x-ray diffraction. Examples of this are x-ray correlation lengths of 250Å in lead strearate (*6,7*), 31Å in arachidic acid (*8*) and 500Å in palmitoyl-R-lysine (*9*). Molecular dynamic simulations (*10-12*) and crystal nucleation theory (*13*) also predict phase coexistence in monolayers on water surfaces.

In a thorough description of monolayers at the air-water interface, the monolayer thickness and the chain orientation are defined. Kjaer et al. (*8*) used the peak position in a synchrotron x-ray reflectivity experiment to measure the monolayer thickness. The phospholipid dimyristoylphosphatidic acid was found to increase in thickness by 40% during the steep part of the surface pressure-area curve. This can be the result of crystalline island coalescence or orientation changes. Schlossman et al. (*14*) used a model based on similar electron densities between methylene chains and the water subphase to predict the thickness and chain orientation in lignoceric acid. They demonstrated that the thickness does not change much in the liquid regions, but substantially increases in the condensed region. Using the assumption that the methylene chains are all-*trans* in conformation, molecular tilts were found to be ~30° in the low surface pressure region and 0° in the condensed region, i.e., large changes in orientation occur in the steep part of the surface pressure-area isotherm. On the other hand, the ellipsometric phase angle ($\delta\Delta$) in pentadecanoic acid shows the largest changes during the plateau region (at intermediate molecular areas) (*15*). $\delta\Delta$ increases by about 100% and then levels off as the "knee" in the surface pressure-area curve is approached at low molecular areas. The ellipsometry results can be explained in terms of the sensitivity of $\delta\Delta$ to other variables in addition to the thickness.

Direct orientation measurements of the methylene tail in amphiphilic monolayers are limited and have only been accomplished with vibrational spectroscopy. Sum-frequency vibrational spectroscopy (SFG) is surface specific and suitable for amphiphilic monolayer studies. Conformation and orientation information may be obtained from the peak intensities. SFG studies of pentadecanoic acid indicate that the methylene chain axes are located within 10° of the surface normal at high surface pressure values (16). At low surface pressures, the appearance of methylene stretching vibrations indicates a breakdown in the SFG selection rules and the introduction of gauche structures.

Information concerning the orientation of polar head groups in amphiphilic monolayers on aqueous substrates has been obtained with optical second harmonic generation (17-20) and is often correlated with surface pressure-area isotherms. From the orientation of the head groups, molecules of pentadecanoic acid are found to stand perpendicular to the interface at high surface pressure. At low surface pressure the hydroxyl bond in the acid group becomes perpendicular to the interface (17). Break points in orientation angle-surface area curves occur at the onset of the plateau region (gas to liquid-expanded phase transition). In amphiphiles containing a naphthyl sulfonate head group, a good correspondence is observed between a featureless surface pressure-area curve and the absence of an abrupt change in orientation (18).

The hydrophobic part of surfactant molecules is often composed of methylene chains. The conformational state of the chains is of importance because molecular models that are used to predict molecular structure are usually based on the assumption that the chains are all-*trans* in conformation. Changes in conformation can easily be followed with the methylene stretching vibrational peaks (21). The conformational state of monolayers on aqueous substrates has been explored by Dluhy, Mendelsohn, et al. with reflection-absorption infrared spectroscopy (22,23). In the phospholipid 1,2-dipalmitoyl-*sn*-glycero-3-phosphocholine, gradual changes in the frequency of methylene stretching modes were found to accompany abrupt surface pressure-area changes. Similar studies on an extracted pulmonary surfactant also showed gradual frequency changes as the molecular area is reduced. This behavior is in contrast to temperature-induced phase transitions in bulk aqueous solutions of phospholipids where abrupt frequency changes occur in response to a conformational randomization on melting (24).

Optical Models in the Literature

Optical models of the interaction of polarized radiation with plane parallel layers are based on Maxwell's equations. Many derivations have appeared in the literature. The most generalized models treat the probed material as an *n*-layered system with an arbitrary variation in optical properties with distance. Optical modeling was first applied to an isotropic monolayer deposited on an aqueous subphase by Dluhy (25). In this work, it was established that when the optical constants of the substrate are lower than the monolayer, non-zero electric field intensities exist in three dimensions, unlike the cases of grazing-angle reflection from metal surfaces or "normal" transmission spectroscopy.

An extension of optical treatments to include anisotropy in the optical constants provides a means to estimate chain orientation and surface concentration from reflection-absorption intensities. Models of this kind are based on a definition of optical constants as in Figure 1. The ambient superphase and liquid subphase are isotropic, and the monolayer has the indicated anisotropic optical constants. Fina and Tung (*26*) originally used such a model to predict the dependence of the reflection-absorption of a monolayer on the chain orientation. The reflected amplitudes for a three phase system are found from the well known relationship:

$$r = \frac{r_1 + r_2 \exp[-2i(a - ib)\eta]}{1 + r_1 r_2 \exp[-2i(a - ib)\eta]}$$

with $\eta = 2\pi(d/\lambda)$, d = monolayer thickness, λ = wavelength, and a and b are phase constants. In the above equations r_1 and r_2 are the reflected amplitudes from each interface and can be written in terms of the anisotropy in the complex refractive index. The effect of chain orientation on the s- and p-polarized reflection-absorption is shown in Figure 2, where RA = -log(R/R$_o$) and the chain orientation angle is defined as the tilt angle from the surface normal. The parameters used in the model calculations are: $\theta = 30°$, n(H$_2$O) = 1.415 and k(H$_2$O) = 0.0163 at 2920 cm^{-1}, a transition dipole moment directed perpendicular to the all-*trans* methylene chain with complete rotation about the chain axis, a k$_{max}$ a value of 0.30 where k$_{max}$= 3k$_{isotropic}$, a birefringence in the infrared frequency range of 0.055, and a monolayer thickness that depends only on orientation (d=17.8Å when chains are perpendicular to the interface and all-*trans*). The Figure illustrates that p-polarized intensities are always larger than s- and that the intensities are more sensitive to orientation at intermediate chain tilt angles. It has recently been demonstrated that both the chain orientation and the surface concentration can be determined from s- and p-polarized intensities (*27*).

Three additional optical models of the reflection-absorption of an anisotropic monolayer on an aqueous subphase have been developed very recently. Gericke et al. (*28*) derived expressions which include an integration of the depth gradient in the optical constants. They found an excellent match between model intensity predictions and experimental intensities using the angle of incident light as a variable. Hsu (*29*) developed equations to determine the transition dipole moment direction and the absorption coefficient from s- and p-polarized intensities. Buontempo and Rice (*30*) include unrestricted rotation around the chain axis in their model and derive relatively simple expressions for the normal and transverse components of the absorption coefficient. Additionally, using the assumption that $n^2 >> k^2$ for normal components, a straight forward relationship using the dichroic ratio for chain tilt calculations was developed.

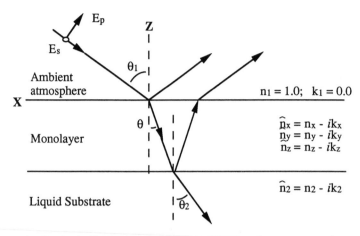

Figure 1. Optical diagram of an oriented monolayer on an isotropic liquid substrate.

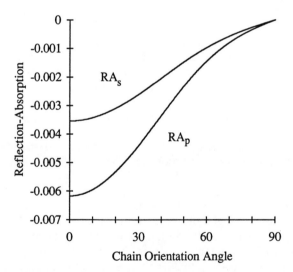

Figure 2. Calculated reflection-absorption (RA) signal intensities for s- and p-polarized light.

In the remainder of this chapter, experimental work on the soluble surfactant sodium dodecyl sulfonate is presented along with suggested interpretations of the peak intensities and frequencies. In addition, the infrared results are compared with a surface tension analysis.

Experimental

Solutions used to form monolayers are prepared from sodium dodecyl sulfonate ($C_{12}S$) and highly purified water. The solutions are placed in a thoroughly cleaned Teflon® trough. Monolayers are formed by self-assembly at the air-water interface. A time of 30 minutes was allowed for an adsorption equilibrium to establish. The surface was scanned with a Digilab FTS60A spectrometer bench and a Specac monolayer/grazing angle accessory. The monolayer trough was isolated vibrationally from the motion of the interferometer and from building vibrations. Spectra are collected by coadding 4096 scans at a resolution of 8 cm^{-1} with a narrow band HgCdTe detector. One level of zero filling and triangular apodization are applied in the Fourier domain.

Results and Discussion

The properties of soluble surfactant monolayers are often characterized with surface tension isotherms, where surface tension is plotted with the bulk concentration (or -log C). The surface tension isotherms display *three* regions of interest in terms of structural considerations presented later. An abrupt slope change occurs at the critical micelle concentration (CMC). At bulk concentrations higher than the CMC (Region 1), the surface tension is constant, as is the surface concentration. At bulk concentration just below the CMC (Region 2), the surface tension rises linearly (in a -log C plot). The surface concentration is also constant in this region. At very dilute bulk concentrations (Region 3), the surface tension tails off to the value of pure water (~72 dynes/cm) and the surface concentration goes to zero. For a thorough treatise in this area, refer to Rosen (*31*). A large variety of studies are present in the literature describing the surface tension and associated thermodynamic properties. Structural inferences have been made from these studies. Nonetheless, very little direct structural information is available for soluble surfactant monolayer systems. With the use of reflection-absorption infrared (RA-IR) spectroscopy, new information is emerging.

One of the key issues in the use of RA-IR spectroscopy for monolayers on aqueous substrates is the formation of an equilibrium state in terms of the adsorption and the structure. The high sensitivity of infrared spectroscopy to concentration, conformation and orientation requires that these variables be constant with time. In this work, the attainment of "vibrational" equilibrium is monitored by the time dependence of the frequencies and intensities of the methylene stretching infrared peaks of

$C_{12}S$. The starting point for the equilibrium process is taken as the time when the well-mixed surfactant solution is poured into the trough. The surfactant molecules immediately begin to adsorb at the air-water interface and structurally rearrange. The smallest time period that can be used in the RA-IR experiment is somewhat less than 2 minutes, the limitation being due to the time required to collect enough scans to attain a reasonable signal-to-noise ratio. Using this time frame, $C_{12}S$ at a bulk concentration of 8.76×10^{-3} M, which is very close to the bulk CMC, shows no time dependence of the frequencies or intensities of the symmetric and asymmetric methylene stretching peaks. This provides evidence that the surface concentration, alkyl chain orientation and alkyl chain conformation are constant and that an equilibrium state is established within the first two minutes. However, a remote possibility exists that these properties change on a longer time scale and that the constant RA-IR signal is the result of a cancellation or combination of effects. At bulk concentrations below the CMC, but still in the linear region (Region 2), a time dependence to the frequencies and intensities is displayed. In about 20 minutes the frequencies decrease 2-5 cm^{-1} and the intensities increase two to three times the value of the initial two minute scan. The interpretation of the decrease in frequency is unambiguous, i.e., the molecules first adsorb to the interface in a disordered conformational state and the order is improved with time. The time dependence of the intensity is related to the increase in conformational order through the larger absorption coefficient of *trans* conformers, as compared to *gauche*. The time dependence can also be functions of the surfactant adsorption and the chain orientation. It is informative to note that the intensity-time profile of the asymmetric methylene stretching peak is accurately fitted with a Langmuir adsorption model implying the importance of surface density in the time dependence of the intensity profile.

RA-IR spectra have been collected from $C_{12}S$ monolayers in the equilibrium state as a function of bulk concentration. The spectra are shown in Figure 3 in the methylene stretching region. The bulk concentrations from bottom to top are 1×10^{-2}, 5.60×10^{-3}, 3.58×10^{-3}, 2.28×10^{-3}, 1.49×10^{-3}, and 1.18×10^{-3} M. The first three bulk concentrations are below the CMC of $C_{12}S$ and in the linear region of the surface tension-log C isotherm (Region 2). The last three concentrations are below the linear region (Region 3). Upon close examination of Figure 3, it is noted that the bottom three spectra (Region 2) are quite similar in intensity and frequency. This provides evidence for the constancy of the surface concentration and structure. Many studies in the literature have demonstrated the constancy of surface concentration in Region 2. The average frequencies of the asymmetric and symmetric methylene stretching peaks in Region 2 are 2917.2 ± 0.2 and 2849.2 ± 0.2 cm^{-1}, respectively. *S*- and *p*-polarized light show no frequency differences. In a previous work it was determined that frequency shifting and band distortion due to the experimental optics are negligible as compared to the transmission mode for

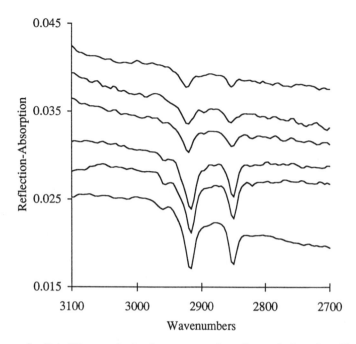

Figure 3. RA-IR *p*-polarized spectra of sodium dodecyl sulfonate monolayers on a water substrate. Bulk Concentrations from bottom to top: 1×10^{-2}, 5.60×10^{-3}, 3.58×10^{-3}, 2.28×10^{-3}, 1.49×10^{-3}, 1.18×10^{-3} M.

the presently used optical constants (*27*). For purposes of comparison, we determined the transmission mode peak frequencies of $C_{12}S$ which had been solvent-cast from an ethanol solution (i.e., a highly crystalline sample). The methylene stretching frequencies are 2917 ± 0.2 (asymmetric) and 2849.6 ± 0.1 cm^{-1} (symmetric). The negligible differences in frequency between the solvent-cast crystals and the monolayer establish that the alkyl chains are in an all-*trans* state in the monolayer in Region 2 of the surface tension isotherm.

An abrupt change occurs between the middle two spectra in Figure 3. The peak intensities drop and close examination reveals that the peak frequencies shift up by 2-3 cm^{-1}. The alkyl chains undergo a cooperative change in ordering. The bulk concentration at the point of abrupt peak change corresponds to the junction of Regions 2 and 3 of the surface tension isotherm. We therefore can define the spectral changes as associated with a phase transition. Further support for this conjecture comes from 1) the fact that surface concentration changes that occur in the bulk concentration range under consideration are much smaller than the intensity decreases in Figure 3 and 2) surface tension versus bulk concentration measurements show an abrupt slope change at the bulk concentration where the RA intensities drop. At the lowest bulk concentrations of Figure 3 (Region 3), the frequencies and intensities suggest that the methylene tails are disordered due to the introduction of *gauche* conformers. s-polarized RA-IR spectra (not shown) show the same behavior and trends as discussed above for p-polarized.

The intensity of s- and p-polarized peaks can be used to calculate the average orientation angle and the absorption coefficient of a specific absorption peak using an optical model as referred to in an earlier section. However, the model can only be used unambiguously when the assumptions are met. Complete rotation about the chain axis must exist. This has been found to be the case in other monolayer-water systems (*32-34*). The alkyl chains must also exist in an all-*trans* conformation. This has been demonstrated in the present work for high bulk concentrations. When the chains are all-*trans*, the monolayer thickness depends only on the chain tilt angle. The orientation of the all-*trans* alkyl tails and the absorption coefficients of the methylene stretching modes are found by independently calculating the parameters for six bulk concentrations and three time trials (after equilibrium is established) within each concentration in the linear region of the surface tension isotherm (Region 2). The results show negligible differences between trials and between concentrations thus establishing the constancy of structure. When the 18 sets are treated as a group, the average chain tilt angle (from the surface normal) is found to be $39.8\pm0.97°$ and the average k_{max} value for the asymmetric methylene stretch (~2917 cm^{-1}) is 0.544 ± 0.014. The angle is in good agreement with that found in a hydrated crystal of $C_{12}S$, i.e. $40°$ (*35*). The k_{max} value can be converted to a surface concentration (Γ) for the purpose of comparison with the Γ value predicted from a surface tension analysis, the latter of which is known to be an accurate measure. Following a procedure described elsewhere (*27*), the k_{max} value of $C_{12}S$ is found to be 1.026 for a crystalline sample that was solvent cast onto a salt window. Using the ratio of k_{max} values as equivalent to the ratio of concentrations in the monolayer and the solid crystal, the two dimensional concentration of $C_{12}S$ monolayers is found to be 3.28×10^{-10} mol/cm^2. This value is in good agreement with that found in a surface tension analysis, 2.90×10^{-10} mol/cm^2 (*36*).

In the lowest bulk concentration region of the surface tension isotherm (Region 3), *gauche* conformers are introduced and therefore the transition dipole moments of the methylene chain tend towards a randomized state. A deviation from the all-*trans* conformational state limits the applicability of the optical model for chain tilt determinations. Keeping this in mind, the optical model calculations correctly predict a substantial drop in the k_{max} value just below the phase transition, which is due to the introduction of *gauche* conformers. However, contrary to expected behavior, the k_{max} values do not decrease further as the bulk concentration is lowered. To address this problem, the chain tilt angle is fixed at the random angle of $54.74°$ in the calculations. Whereupon, the k_{max} values decrease monotonically with bulk concentration, as expected. However, the k_{max} values do not significantly change through the phase transition. The opposing nature of the two calculations suggest that the alkyl chains become partially randomized at bulk concentration just below the phase transition. The SO_3 stretching region of the RA-IR spectra (1300-1000 cm^{-1}) provides information about the head group orientation and structure. The signal-to-noise in this region is not as good as in the methylene stretching region, so qualitative considerations are presented. Using the assumption that the SO_3 symmetric stretching vibrational mode is directed parallel to the all-*trans* alkyl chain, and therefore, the asymmetric mode to perpendicular, orientation changes of the head group can be followed. An increase in the asymmetric to symmetric intensity ratio indicates that the direction perpendicular to the plane containing the three oxygen atoms moves closer to the surface normal. This is precisely what is observed when salt ions are added in excess to the water subphase. The cations of the salt interact with the head group and change the head group orientation. This conclusion is based on a consideration of the effect of salt ions on the absorption coefficients of the sulfonate stretching modes, which will be presented by the authors in more detail elsewhere. The change in orientation of the head groups in the presence of salt ions is found to reduce the order in the methylene chains.

Literature Cited

(1) Kjaer, K.; Als-Nielsen, J.; Helm, C. A.; Laxhuber, L. A.; Möhwald, H. *Phys. Rev. Lett.* **1987**, *58*, 2224.

(2) Helm, C. A.; Möhwald, H.; Kjaer, K.; Als-Nielsen, J. *Biophys. J.* **1987**, *52*, 381.

(3) Möhwald, H. *Thin Solid Films* **1988**, *159*, 1.

(4) Rasing, T.; Hsiung, H.; Shen, Y. R.; Kim, M. W. *Phys. Rev. A* **1988**, *37*, 2732.

(5) Pallas, N. R.; Pethica, B. A. *Langmuir* **1985**, *1*, 509.

(6) Lin, B.; Peng, P.; Ketterson, J. B.; Dutta, P. *Thin Solid Films* **1988**, *159*, 111.

(7) Dutta, P.; Peng, J. B.; Lin, B.; Ketterson, J. B.; Prakash, M.; Georgopoulos, P.; Ehrlich, S. *Phys. Rev. Lett.* **1987**, *58*, 2228.

(8) Kjaer, K.; Als-Nielsen, J.; Helm, C. A.; Tippmann-Krayer, P.; Möhwald, H. *Thin Solid Films* **1988**, *159*, 17.

(9) Wolf, S. G.; Landau, E. M.; Lahav, M.; Leiserowitz, L.; Deutsch, M.; Kjaer, K.; Als-Nielsen, J. *Thin Solid Films* **1988**, *159*, 29.

(10) Mann, J. A.; Tjatjopoulos, G. A.; Azzam, M. J.; Boggs, K. E.; Robinson, K. M.; Sanders, J. N. *Thin Solid Films* **1987**, *152*, 29.

(11) Shin, S.; Collazo, N.; Rice, S. A. *J. Chem. Phys.* **1992**, *96*, 1352.

(12) Shin, S.; Collazo, N.; Rice, S. A. *J. Chem. Phys.* **1993**, *98*, 3469.

(13) Lando, J. B.; Mann, J. A. *Langmuir* **1990**, *6*, 293.

(14) Schlossman, M. L.; Schwartz, D. K.; Kawamoto, E. H.; Kellogg, G. J.; Pershan, P. S.; Ocko, B. M.; Kim, M. W.; Chung, T. C. In *Materials Research Society Symposium Proceedings, Macromolecular Liquids*; C. R. Safinya, S. A. Safran and P. A. Pincus, Ed.; Materials Research Society: Pittsburgh, 1990; Vol. 177; pp 351-361.

(15) Kim, M. W.; Sauer, B. B.; Yu, H.; Yazdanian, M.; Zografi, G. In *Materials Research Society Symposium Proceedings, Macromolecular Liquids*; C. R. Safinya, S. A. Safran and P. A. Pincus, Ed.; Materials Research Society: Pittsburgh, 1990; Vol. 177; pp 405-410.

(16) Sionnest, P. G.; Hunt, J. H.; Shen, Y. R. *Phys. Rev. Lett.* **1987**, *59*, 1597.

(17) Rasing, T.; Shen, Y. R.; Kim, M. W.; Grubb, S. G. *Phys. Rev. Lett.* **1985**, *55*, 2903.

(18) Rasing, T.; Shen, Y. R.; Kim, M. W.; Valint, J., P.; Bock, J. *Phys. Rev. A* **1985**, *31*, 537.

(19) Vogel, V.; Mullin, C. S.; Shen, Y. R.; Kim, M. W. *J. Chem. Phys.* **1991**, *95*, 4620.

(20) Grubb, S. G.; Kim, M. W.; Rasing, T.; Shen, Y. R. *Langmuir* **1988**, *4*, 452.

(21) Snyder, R. G.; Hsu, S. L.; Krimm, S. *Spectrochim. Acta* **1978**, *34A*, 395.

(22) Mitchell, M. L.; Dluhy, R. A. *J. Am. Chem. Soc.* **1988**, *110*, 712.

(23) Dluhy, R. A.; Reilly, K. E.; Hunt, R. D.; Mitchell, M. L.; Mautone, A. J.; Mendelsohn, R. *Biophys. J.* **1989**, *56*, 1173.

(24) Pastrana, B.; Mautone, A. J.; Mendelsohn, R. *Biochem.* **1991**, *30*, 10058.

(25) Dluhy, R. A. *J. Phys. Chem.* **1986**, *90*, 1373.

(26) Fina, L. J.; Tung, Y. S. *Appl. Spectrosc.* **1991**, *46*, 986.

(27) Tung, Y. S.; Gao, T.; Rosen, M. J.; Valentini, J. E.; Fina, L. J. *Appl. Spectrosc.* **1993**, *47*, 1643.

(28) Gericke, A.; Michailov, A. V.; Hühnerfuss, H. *Vib. Spectrosc.* **1993**, *4*, 335.

(29) Hsu, S. L. *personal comm.* **1994**.

(30) Buontempo, J. T.; Rice, S. A. *J. Chem. Phys.* **1993**, *98*, 5825.

(31) Rosen, M. J. *Surfactants and Interfacial Phenomena;* 2nd ed.; John Wiley & Sons: New York, 1989.

(32) Paudler, M.; Ruths, J.; Riegler, H. *Langmuir* **1992**, *8*, 184.

(33) Ducharme, D.; Max, J.-J.; Salesse, C.; Lablanc, R. M. *J. Phys. Chem.* **1990**, *94*, 1925.

(34) Bohanon, T. M.; Lin, B.; Shih, M. C.; Ice, G. E.; Dutta, P. *Phys. Rev. B* **1990**, *41*, 4846.

(35) Lingafelter, E. C.; Jensen, L. H. *Acta Cryst.* **1950**, *3*, 257.

(36) Dahanayake, M.; Cohen, A. W.; Rosen, M. J. *J. Phys. Chem.* **1986**, *90*, 2413.

RECEIVED September 6, 1994

Chapter 5

A New Generation of Mass Spectrometry for Characterizing Polymers and Related Materials

Long-Sheng Sheng[1], Sanford L. Shew, Brian E. Winger, and Joseph E. Campana[2]

Extrel FTMS Inc., Waters,
6416 Schroeder Road, Madison, WI 53711

Mass spectrometry is a powerful analytical technique for characterizing polymers. New mass spectrometry ionization methods have extended the molecular weight range over which macromolecules can be ionized to millions of Daltons. And new breeds of mass spectrometer analyzers and detectors are quickly extending the absolute measurable molecular weight range into the millions. More important to polymer scientists and engineers than these high mass achievements in mass spectrometry is that today's mass spectrometers can determine structural information on oligomers quickly, directly, and routinely for several classes of compounds. In particular for polymer systems below 10,000 Daltons, few indirect analytical methods can provide the information provided directly by mass spectrometry. Some state-of-the-art mass spectrometry techniques are discussed and several examples are presented to illustrate molecular weight and structural determination by mass spectrometry.

Mass spectrometry is rapidly evolving in its scope of applications for macromolecular analysis through new ionization techniques. For example, laser desorption ionization (LDI) techniques coupled with time-of-flight mass spectrometry (TOFMS) can produce accurate molecular weight information quickly for molecular weights of a quarter million. In a single laser shot, the LDI technique coupled with Fourier Transform Mass Spectrometry (FT/MS) can provide detailed chemical information on polymeric molecular structure and provide direct determination of additives and contaminants in polymers. State-of-the-art mass spectrometry methods can be coupled with gel permeation chromatography (GPC) in an off-line mode, or directly coupled via electrospray ionization for the analysis of macromolecules. All of these mass spectrometry techniques offer new analytical capabilities to solve problems in research, development, engineering, production, technical support, competitor product analysis, and defect analysis.

[1]Visiting Scientist on sabbatical leave from the China Pharmaceutical University, Nanjing, China
[2]Corresponding author

Mass Spectrometry Today

During the last 10 years, the growth in mass spectrometry applications in the biosciences has been remarkable. There have been a number of advances in mass spectrometry instrumentation that have allowed the molecular weight and structure of peptides and proteins to be determined. Today, direct protein analysis is performed by mass spectrometry with matrix-assisted laser desorption ionization (*1, 2*) and electrospray ionization [ESI] (*3*). On-line liquid chromatography mass spectrometry (LC/MS) of proteins is becoming commonplace. In addition, mass spectrometers have evolved from finicky research instruments to routine user-friendly shared spectrometers. The same advantages that mass spectrometry is providing to the biosciences may be realized in polymer science.

Laser Desorption/Ablation Ionization Methods. Matrix-assisted laser desorption ionization (MALDI) relies on the use of a solid chromophoric matrix, chosen to absorb laser light, which is co-mixed with the analyte (*4*). Typically, a solution of a few picomoles of analyte is mixed with a 100-to-5000 fold excess of the matrix in solution. A few microliters of the resulting solution are deposited on a mass spectrometer solids probe, and the solvent is allowed to evaporate before inserting the probe into the mass spectrometer. When the pulsed laser beam strikes the sample surface in the spectrometer, the sample molecules are desorbed/ionized at high efficiency. The various mechanisms for matrix and non-matrix assisted laser desorption have been discussed (*5, 6*).

A matrix is not always necessary or desirable to obtain a mass spectrum of a sample. For example, to rapidly determine if an ultraviolet (UV) absorbing additive is present in a paint sample by LDI mass spectrometry, it is not convenient to alter the sample to introduce a matrix. Instead, a paint chip could be directly analyzed by laser desorption with a UV or other type of laser.

Pulsed lasers with relatively short pulse widths (<50 ns) are typically used for laser desorption/ablation techniques because of their high peak powers and because short pulses reduce sample consumption and minimize laser-induced pyrolysis of the sample. Only mass spectrometers that can measure ions of all mass-to-charge values near simultaneously (corresponding to a single short laser pulse) or mass spectrometers that can trap all the ions produced in a single laser pulse are compatible with this pulsed ionization technique. Time-of-flight (*2,7-9*) and Fourier transform mass spectrometers (*9-14*) are commercially proven for laser desorption/ablation mass spectrometry. The TOFMS detects all ions near simultaneously, and the FT/MS is a ion trapping mass spectrometer that detects all ions simultaneously.

Other types of mass spectrometers, low, medium, or high performance, which scan by measuring one mass-to-charge ratio at a time cannot be used effectively for laser desorption (*9*). High-resolution magnetic sector mass spectrometers, fitted with array detectors to allow simultaneous ion detection, have been shown to be able to measure laser desorption mass spectra over a limited mass range and with relatively low resolving power (*15*). Quadrupole ion trap mass spectrometers (radiofrequency ion traps) capture and store all the ions formed in a single laser burst; however, they measure ions sequentially over relatively long periods of time (*16*) and produce low-resolution mass spectra. During the long measurement time the trapped ions may undergo ion chemistry and ion collisions that affect the integrity of the ions that are ultimately measured.

Electrospray Ionization. Electrospray ionization is a recent ionization technique that has been applied to large macromolecules such as proteins, although it is applicable to other large molecules including polymers. It is a method for transforming ions that are present in a solution (*17*) into characteristic ions in the gas phase (mass spectrometers can only analyze ions in the gas phase). A sample solution is sprayed or nebulized under the influence of a high electric field. The resulting aerosol is desolvated by a combination of heat, gas flow, and vacuum; and by using supersonic beam methods, a beam of characteristic gas-phase ions is simply and efficiently formed for mass spectrometric analysis. The electrospray ionization method often results in the formation of multiply-charged molecules in the gas phase for high molecular weight species.

With electrospray, molecules are produced with a distribution of charge states. In other words, positively-charged macromolecules may be produced with, for example, a distribution of 10, 11, 12, etc. protons (or other cations) attached. Therefore, the mass-to-charge ratio of the multiply-charged macromolecule is some fraction of the molecular weight, for example, a tenth, eleventh, twelfth, etc., which lowers the mass-to-charge range at which the ion distribution resides. However, this complicates the determination of the true mass of the macromolecule. With conventional low-resolution mass spectrometers, the true mass of the macromolecule is determined by an indirect and iterative computational method.

With the Fourier transform mass spectrometry technique, the ^{13}C isotopes corresponding to each charge state can be easily resolved in the mass spectrum, which cannot be accomplished with lower resolution mass spectrometers. By counting the number of ^{13}C isotopic peaks in a single mass unit, or equivalently by dividing the mass difference between two adjacent ^{13}C peaks in a given charge state's isotopic cluster into unity gives the charge-state value by direct measurement (*18*). Therefore, the mass difference between the ^{13}C peaks allows the molecular weight to be directly determined. In other words, the molecular weight is calculated by multiplying the apparent mass-to-charge ratio by the charge state less the mass of the attached cations.

Electrospray ionization has three advantages for mass spectrometry. First, it allows large macromolecules to be ionized, and second, because the ions formed by ESI are typically multiply charged, the mass-to-charge ratios of ionic macromolecules are decreased into a mass regime where mass spectrometers operate most effectively. For example, polymers up to 5,000,000 have been measured by electrospray ionization mass spectrometry (*19*), and more recently single ions of several million in molecular weight have been measured by FT/MS (*20*). Third, because the input to electrospray ionization is a liquid, it serves as an interface between the mass spectrometer and liquid chromatographic techniques including gel permeation chromatography (GPC) also called size-exclusion chromatography [SEC] (*21*), and capillary electrophoresis (*22*).

Determination of Molecular Weight Distribution

Why use Mass Spectrometry? Mass spectrometry can provide the most accurate mass determination of all analytical techniques, and furthermore, the measurement is direct. Today, the strength of mass spectrometry lies in determinations below 100,000 Daltons. It allows the direct analysis of solid materials including bulk, surface and additive chemistries. Mass spectrometry has relatively high sensitivity, dynamic range and linearity. Mass spectrometry, unlike chromatography, is comparatively fast and has capabilities to provide structural information. Also, mass spectrometry has the capability to resolve or separate components of complex mixtures on the mass scale.

Why Chromatography? Gel permeation chromatography is accuratè and routine for average molecular weight determinations in excess of 50,000. Chromatographic methods separate complex mixtures resolved in time. For example, in a GPC an oligomer mixture is separated by hydrodynamic size (volume).

The GPC separated mixture can be used as a direct sample input into a mass spectrometer for mass analysis. As will be discussed below, the limitation in GPC is more related to the traditional detectors than the chromatography. The mass spectrometer is a more ideal detector for liquid chromatography.

Gel Permeation Chromatography. The use of GPC in combination with viscometry and light scattering detectors for the determination of a polymer's absolute molecular weight distribution is becoming more common. However, there can be limitations of these analytical systems for the analysis of low molecular weight polymers (<50,000 Daltons). Viscometry and light scattering detectors are not as sensitive to lower molecular weight polymers under normal GPC conditions. The concentration detector is usually a refractometer that may not always give reliable concentration information below 50,000 Daltons without meticulous calibration. Yet, many industrial laboratories routinely rely on GPC measurements below 50,000 Daltons and even below 10,000 Daltons.

The viscometer sensitivity is dependent upon the specific viscosity of the sample so it is not uncommon for an analyst to compensate by using high concentrations of low molecular weight samples to get reasonable signal-to-noise. These high concentrations in turn influence the elution time (volume) of the unknown polymer sample. An accurate elution volume is critical for the molecular weight determination by using the universal GPC calibration.

A light scattering detector's sensitivity is a function of molecular weight. With polymers below 25,000 Daltons, the sensitivity is a limitation. The signal from the light scattering detector is due to the sample's excess Rayleigh factor R_0. The excess Rayleigh factor is a function of the square of the change in refractive index with concentration $[(dn/dc)^2]$. Therefore, if the polymer exhibits a low dn/dc, less than 0.06 under normal GPC concentrations, the detector may not provide sufficient signal-to-noise to calculate the molecular weight distribution.

It is well known that the refractive indices of low molecular weight polymers in solutions are affected adversely by different end groups. In other words, at low molecular weight, the detector is more sensitive to chemical composition than it is to molecular weight. The use of a refractive index detector for low molecular weight concentration determinations can be unreliable because the dn/dc of the sample may change as a function of molecular weights below 50,000 Daltons.

Mass Spectrometry. An alternative to using GPC with viscometry and light scattering detection for the determination of low molecular weight polymer distributions is mass spectrometry. Matrix-assisted laser desorption ionization time-of-flight mass spectrometry (MALDI TOFMS) can give fast, accurate information about the degree of polymerization (molecular weight distribution).

Figure 1 compares a gel permeation chromatogram, a MALDI TOFMS mass spectrum, and a MALDI FT/MS mass spectrum of an unknown industrial polymer. The mass spectra were obtained by using 2,5-dihydroxybenzoic acid (DHBA) as the matrix and 337 nm radiation from a nitrogen laser. The MALDI TOFMS result was obtained by averaging over 50 laser shots. The MALDI FT/MS result was obtained by measuring a single laser shot, which is indicative of higher sensitivity.

The results from both MALDI TOFMS and Laser Probe FT/MS illustrate higher molecular weight and higher resolution data than that obtained by GPC. In addition, the higher resolving power achievable with the FT/MS instrument, compared to either GPC and TOFMS, demonstrates that each individual oligomer is resolved on the mass scale. Furthermore, it illustrates how molecular information can be obtained from a single mass spectrum. For example, the accurate mass FT/MS spectrum establishes unequivocally that each oligomer measured has a mass difference of 44.026 which corresponds to C_2H_4O -- the repeating unit of

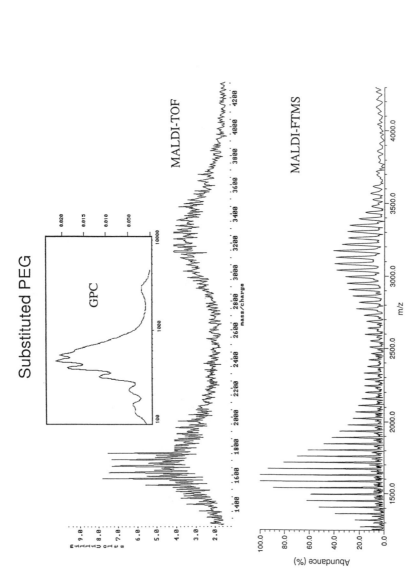

Figure 1. Comparison of analytical results obtained on an industrial polymer by gel permeation chromatography (top) and matrix-assisted laser desorption ionization mass spectrometry (linear time-of-flight mass spectrometer [center] and Fourier transform mass spectrometer [bottom]). From the FT/MS data the polymer can be identified as a substituted polyethylene glycol.

polyethylene glycol. This data establishes that the polymer is some type of polyethylene glycol, whereas the GPC and TOFMS data only give information about the molecular weight. The GPC data illustrates the problem of that method when analyzing unknown and/or low molecular weight polymers even though in this case the GPC experts in our company calibrated and acquired the data. The two mass spectrometric methods are in excellent qualitative agreement; however, the GPC data severely underestimates the molecular weight.

Figure 2 illustrates the application of Laser Probe FT/MS to monitoring a feedstock. In this example, a batch of poly(dimethylsiloxane) was specified to have an average molecular weight of approximately 2,000 Daltons. The direct analysis of the polymer by Laser Probe FT/MS reveals a bimodal distribution between 2,400 and 8,000 Daltons. These results were obtained by a traditional laser desorption ionization method by using a carbon dioxide laser operated at 10.6 µm and by doping potassium bromide into the sample. Alkali halide doping is a common method used with infrared lasers to assist ionization by alkali metal attachment. The series of ionized oligomers results from attachment of the potassium ion to each oligomer (M_n+K^+). From these data, the mass spectrometer data station extracts relative oligomer abundances and computes a weight average molecular weight of 5,300. In this example, the polymer average molecular weight was quite different than what the manufacturer had thought was produced and sold.

The first two examples illustrate bimodal distributions, which can often be characteristic of real industrial samples. The next example (Figure 3) shows the results obtained from a commercial reagent sample of polymethyphenylsiloxane. These results were obtained by matrix-assisted laser desorption ionization (MALDI) technique where ultraviolet radiation, typically from a nitrogen laser (337 nm) is used with a UV absorbing matrix (in this case, dihydroxybenzoic acid [DHBA]) that is co-mixed with the polymer sample. The mass spectrum illustrated was obtained by a single laser shot. As can be seen in the figure, a distribution centered about m/z 3,000 is observed. The molecular weight given on the bottle, which was likely determined by a traditional and indirect method, was 2,600.

Compared to other ionization methods, laser desorption data are not only fast and easy to obtain, but they also appear to yield the highest average molecular weight. This may be explained by the harshness of the other ionization techniques that tear apart the molecular entities during the ionization process. Figure 4 is a Laser Probe FT/MS spectrum of Oxypruf-20 [tetrakis(hydroxypropoxypropoxy-propoxypropoxypropyl)]hydrazine obtained by direct carbon dioxide (10.6µm) laser desorption. In this case, the distribution shows an abundant series of $[M+K]^+$ peaks and to a lesser extent $[M+Na]^+$ peaks (salts were not added; the alkali metals are ubiquitous in many materials). The peaks in each series are separated by 58 Daltons, which is the mass of the repeating group. The weight average molecular weight determined in our laboratory was 1,313.3. In a separate published study (23), the weight average molecular weight was determined to be 1,314.6 by laser desorption FT/MS. Secondary ion mass spectrometry (SIMS) and fast atom bombardment (FAB) ionization techniques, gave values of 773.7 and 569.2 respectively, which illustrates that they may not be as generally well suited for molecular weight determinations.

Table I lists the various classes of polymers that have been studied by laser desorption ionization FT/MS techniques. Molecular weight information or structural information or both has been obtained on all of the polymers listed. In many cases, derivatives, homologous series, or different molecular weight distributions within each class listed have been reported.

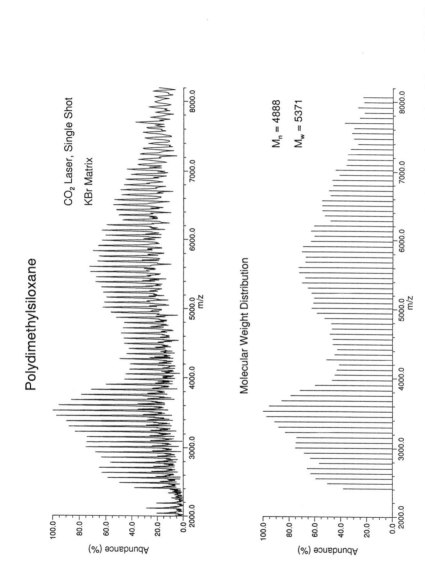

Figure 2. A Laser Probe FT/MS spectrum (top) of an industrial polymer with a theoretical molecular weight of about 3,000. The mass spectrum illustrates a bimodal molecular distribution of poly(dimethylsiloxane) oligomers. The weight average molecular weight calculated from these mass spectrometery data is 5,295 (bottom).

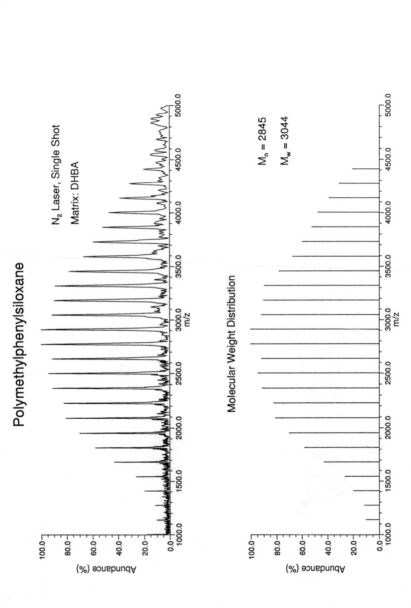

Figure 3. A Laser Probe FT/MS MALDI spectrum (top) of a commercial polymer reagent whose molecular weight is suggested to be 2,600. The weight average molecular weight ($M_W = 3044$) is computed directly from the mass spectrometry data (bottom).

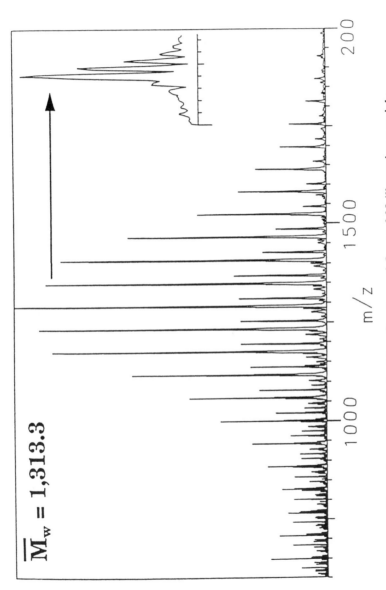

Figure 4. Laser Probe Mass Spectrum of Oxypruf-20 illustrating a weight average molecular weight of 1,313.3. These data are consistent with a published value of 1314.6 which was obtained independently by Laser Probe FT/MS (23).

**Table I.   Classes of Polymers Studied by Laser Desorption
                Fourier Transform Mass Spectrometry[a]**

acrylic acid	octylphenol ethoxylates
acrylonitriles (26)	oxypropylenediamines
alkoxylated pyrazoles (23)	p-phenylenes (29-30, 40, 41)
alkoxylated hydrazines (23)	peptides
amic acid (27)	perfluoro ethers (37, 42)
amides	phenyl-pyrrolylenes (28-30, 39)
analines (28, 29)	phenylene sulfides (25, 26, 28, 29)
brominated p-phenylenes (30)	propylene glycols (32, 34)
butadienes (31)	pyrenes (29)
caprolactone diols (32)	saccharides
dimethylsiloxanes (33)	selenienylenes (28, 30, 39)
diols	styrenes (26, 31, 35)
esters	sulfonic acids
ethylene glycols (25, 32-34)	tetrafluoroethylenes
ethylene glycol methylethers (32)	thienylenes (28-30, 39)
ethylene imines (32)	triols
ethylene terephthalates (35)	vinyl acetates (25)
ethylenes (26, 31)	vinyl phenols (29)
fluorocarbons (36, 37)	vinylchlorides (26)
hydroxybutylic acid	
hydrocarbon waxes	**copolymers**
imides (25, 27, 33)	β-hydroxyalkanoates (esters)
isoprenes (31)	dimethylsiloxane/ethylene oxide
kapton	ethylene glycol/propylene glycol (43)
methyl methacrylates (33, 38)	ethylene/tetrafluoroethylene (25, 26)
methylphenylsiloxanes	methylmethacrylate/butylacrylate (43)
methyl-dipyrrolylenes (28-30, 39)	methylmethacrylate/methacrylic acid
monols	methylmethacrylate/styrene (43)
nucleotides	pyromelletic dianhydride/
nylons	oxydianiline (25)

[a]   Polymers (non-biological) without references are unpublished work from the
      authors' laboratories.

Structural Determination

As mentioned previously, FT/MS is capable of providing more detailed structural information compared to most any other analytical technique. For example, it is possible to measure the absolute mass of an oligomer or a polymer fragment ion, and
from that mass measurement the most probable elemental composition can be computed by the mass spectrometer data station. A more accurate mass measurement
(i.e. a lower error on the measurement) limits the chemical formulas that can be computed. Generally this elemental composition information is all that is needed to confirm a suspected molecular structure or substructure. However, if this is not sufficient, mass spectrometry/mass spectrometry (MS/MS) techniques (24) can be applied.
        Figure 5 conceptualizes how the end groups on the industrial sample of
polyethyleneglycol (PEG) discussed in Figure 1 can be determined by accurate mass
measurement. The mass spectrum and expansion plot shown in Figure 5 were ob-

tained by mixing a PEG standard with the industrial PEG. Two series of ions can be observed and measured with high mass accuracy in the mass spectrum. The series (nominal mass) of higher abundance species (1466, 1510, 1554, ...) result from potassium ion attachment to the PEG standard. The second series (nominal mass) of ions of lower abundance (1524, 1568, 1612, . . .) result from ionization of the industrial PEG. The two series differ by 14 mass units; however, because the measured mass difference is accurate to better than a few parts per million, mass differences attributed to the obvious CH_2, N, etc. can be eliminated.   A further clue about the structure of the endgroup was obtained by doping the sample with KBr. It was observed that the mass of the industrial polymer shifted, compared to the mass determined by the mass spectrum obtained with only the DHBA matrix (Figure 1), corresponding to the displacement of a cation by the potassium ion. The expansion plot in Figure 5 illustrates the molecular formula for the PEG standard and the industrial polymer. It was confirmed that the industrial polymer was synthesized to have at least one anionic terminal group, and these results confirmed the presence of an anionic group by the observation that a proton was displaced and exchanged by the potassium ion after doping with KBr. With some additional information from the polymer chemist, the molecular structure of the end group could be determined by accurately measuring the mass difference between the PEG standard and the industrial PEG sample. The modified PEG end group will have a molecular substructure mass corresponding to one of the following mass differences:

$$[(C_2H_4O)_nX^-K^+]K^+ - (C_2H_4O)_nK^+ + m(C_2H_4O), \text{ where } m = 1, 2, 3, ...$$

This is simply the measured mass difference between two corresponding oligomers of the industrial and standard PEG plus the molecular mass of the repeating unit taken m times. Starting at m=1 and incrementing it, the chemical composition is computed by using the mass spectrometer data station. Choices are limited by specifying that the chemical composition only contains C, H, O, K, and X. At some value of m, a chemically-sensible elemental composition is recognized. A proposed endgroup structure can be further supported by using MS/MS techniques on one of the oligomeric ions and confirmed further by other analytical methods if warranted. (The structure of the endgroup and X cannot be revealed because of the proprietary nature of the sample).

In the MS/MS of polymers, a single oligomeric ionic species is isolated in the mass spectrometer. This is easily accomplished in an FT/MS instrument, which is a magnetic ion trap. Therefore, all the oligomeric species, except the one of interest are ejected from the FT/MS ion trap, hence leaving a population of one oligomer. After this gas-phase isolation is completed by the data station, the selected oligomer is broken down inside the mass spectrometer by using one of several common dissociation techniques. Finally, the dissociation products are measured by acquiring a product ion mass spectrum.

Figure 6 illustrates MS/MS on the n=6 oligomer of a copolymer of β-hydroxyalkanoate, which was determined to contain five C_8 and one C_6 repeating units. After dissociation of the n=6 species, a mass spectrum is measured. Four distinctive fragmentation patterns are observed depending on which end of the oligomer sequentially dissociates. The data illustrate that the oligomer dissociates by four different sequential cleavages. These correspond to sequential loss of three monomeric units:

1. C_8, C_8, C_8
2. C_8, C_8, C_6
3. C_8, C_6, C_8
4. C_6, C_8, C_8.

This illustrates that the C_6 repeating group is randomly incorporated into the n=6 copolymer. This example serves to illustrate that MS/MS is useful for the characterization and determination of the sequence of copolymers.

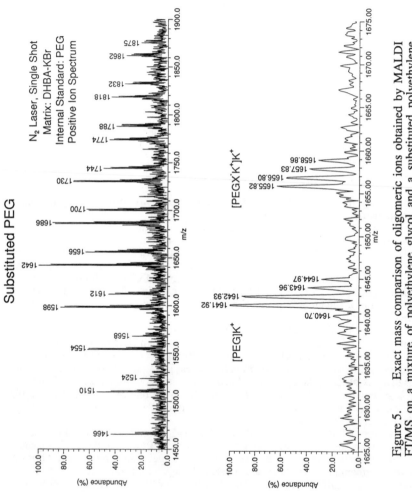

Figure 5.   Exact mass comparison of oligomeric ions obtained by MALDI FT/MS on a mixture of polyethylene glycol and a substituted polyethylene glycol (Same industrial sample as represented in Figure 1.   The end group composition may be determined by these data.)

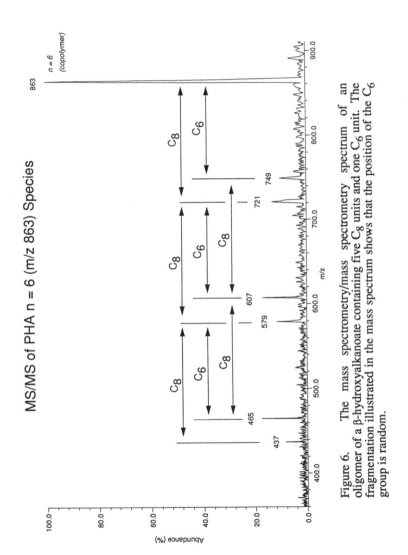

Figure 6. The mass spectrometry/mass spectrometry spectrum of an oligomer of a β-hydroxyalkanoate containing five C_8 units and one C_6 unit. The fragmentation illustrated in the mass spectrum shows that the position of the C_6 group is random.

Additives and Contaminants

The laser-based methods discussed thus far can be used to probe surface and interstitial contaminants as well as for the direct determination of additives in a complex matrix. It is now common to have a CCD camera and video display that gives a microscopic view of the sample when it is in the mass spectrometer. This allows contaminants, defects, and areas of interest to be observed and manipulated while under the probing beam of the laser (25). A number of industrial examples have proved that direct Laser Probe FT/MS analyses can rapidly determine many additives directly, even when the combination of laborious classical wet chemical techniques with other modern instrumental methods have proved difficult and time consuming.

Figure 7 illustrates the direct determination of two antioxidants in a cross-linked polymer. The protonated molecule is observed for Tinuvin 900 and the odd-electron molecular ion is observed for Tinuvin 770. In this example, a carbon dioxide laser was used to ablate the sample, and following the ablation step an electron beam was turned on to ionize the ablated materials. Exact mass measurements confirmed the identity of the two additives present at about 2%. Each laser shot ablates very small quantities of the sample producing a mass spectrum as illustrated in Figure 7. The molecular ion signals observed correspond to the detection of about 30 picomoles of additive in each laser ablation event. In some extreme cases, we have determined detection sensitivities that are sub-attomolar; in particular when determining an UV absorbing surface species with an UV laser probe. Table II illustrates that a wide variety of additives have been determined by Laser Probe FT/MS.

Tinuvin 770

Tinuvin 900

Electrospray and Chromatography

The application of electrospray to polymer characterization is in its infancy. It can be a viable approach to analyzing high molecular weight polymers by taking advantage of the fact that the effective mass-to-charge ratio is decreased into the realm of high-performance mass spectrometry measurements because of the high charge states that are observed.

Figure 8 illustrates a simplistic example of electrospray FT/MS on poly(propylene glycol) [PPG] 720. The solution used to obtain the mass spectrum was approximately 75 micromolar PPG in 1:1 THF:MeOH with 0.5% NaCl. A few picomoles of the PPG was consumed to obtain the mass spectrum. Each oligomer is ionized by sodium ion attachment from the solution. Only the single charge state (singly-charged species) is observed for each oligomer.

As the polymer size increases, the number of charge states will increase for each oligomer. This leads to an extremely complex mass spectrum when there is a distribution of several charge states for each oligomer. Fortunately, the FT/MS is an ultrahigh resolution technique that is capable of resolving the many overlapping charge states of different oligomers. Still, this is may prove to be a formidable computational task.

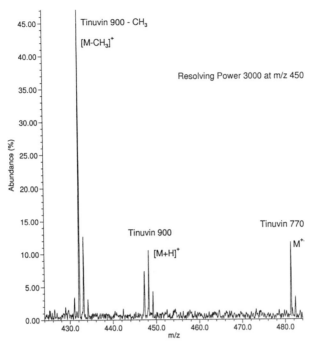

Figure 7.   The Laser Probe FT/MS Spectrum of a piece of a cross-linked polymer illustrating the presence of two antioxidants.

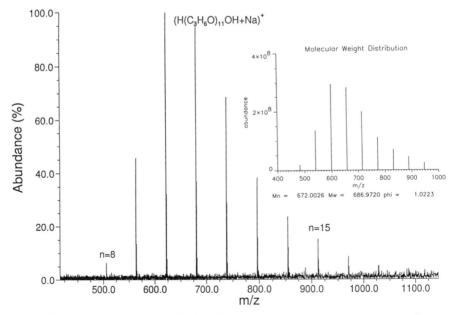

Figure 8.   Electrospray ionization/Fourier transform mass spectrum of poly(propylene glycol) 720. Only the single charge state oligomeric species are observed. The weight average molecular weight calculated from the mass spectrum is 687.

**Table II. Additives Studied by Laser Desorption
Fourier Transform Mass Spectrometry[a]**

Acrawax (44)	Sandostab (47)
Antiozonants (45)	PEPQ
Carbon Black	Seenox (47)
Chimassorb	4125
Dyes (46)	Spinuvex (47)
EBS Wax (44)	A36
Goodrite (47)	Stearamides (44)
3114	surfactants
Irganox (44, 47)	Tinuvin (44, 47)
245, 259, 1010, 1035, 1076,	144, 320, 326, 328, 440,
1098, 3114, MD1024	622, 770
metal stearates (47)	Ultranox (44, 48)
Naugard (44, 47	226, 236, 246, 626
76, 524, BHT, DLTDP, DSTDP	Weston (44)
Oleamides (44)	618, TNNP
Pigments	Wingstay (45)
Polygard (44)	100, 300
	XR 2502 (48)

[a] Additives without references are unpublished work from the authors' laboratories.

An alternative, and a more elegant approach to analyzing large polymers is to couple a GPC via an electrospray ionization interface to the FT/MS. Each oligomeric fraction that elutes from the GPC would be measured directly by mass spectrometry with minimal interference from the multiple charge states of adjacent oligomers.

Summary

Gel permeation chromatography can provide a cost-effective solution to most molecular weight measurements over 10,000 Daltons. Modern laser-based mass spectrometry methods have many applications to polymer characterization. The strengths of mass spectrometry are in the molecular weight regions where GPC performance tails off (<10,000 Daltons) or when detailed molecular structures are needed.

MALDI/TOF provides cost-effective measurements for low molecular weights (less than 100 Daltons) to over 250,000 Daltons. Today, the Laser Probe FT/MS technique can provide accurate molecular weight information for polymers that are less than 20,000 Daltons. Additionally, the FT/MS has features that allow molecular structures and substructures of most classes of polymers to be probed. These mass spectrometry techniques are also directly applicable to contaminant and additive analysis.

There are prospects of coupling GPC to these and other state-of-the-art mass spectrometers. For example, polymer additive analysis can now be performed routinely by LC/MS. Furthermore, there is promise for high-molecular weight characterization (>50,000 Daltons) by using electrospray ionization, another new method for ionizing large molecules in solutions.   The electrospray method is expected to allow GPC to be coupled directly to the FT/MS.

Acknowledgments

The authors thank Richard Nielson and Andy Jarrell (Waters), Trevor Havard (Precision Detectors), William Simonsick, Jr. (E.I. DuPont de Nemours & Co.), Jaap Boon (FOM Institute), James DeVries and Tom Prater (Ford Motor Company) for helpful discussions. The technical assistance of David A. Weil (3M) and Joel Covey (Extrel FTMS, Inc.) for the acquisition of some mass spectra is appreciated.

Literature Cited

1.   Buchanan, M.V.; Hettich, R.L.; *Anal. Chem.* **1993**, *65*, 245A-259A.
2.   Cotter, R.J.; *Anal. Chem.* **1992**, *64*, 1027A-1039A.
3.   Smith, R.D.; Loo, J.A.; Ogorzalek Loo, R.R.;Busman, M.; Udseth, H.R.; *Mass Spectrom. Rev.* **1991**, *10*, 359-451.
4.   Beavis, R.C.; *Organ. Mass Spectrom.* **1992**, *27*, 653-659.
5.   Hillenkamp, F.; Ehring, H.; In *Mass Spectrometry in the Biological Sciences; A Tutorial*; Gross, M.L., Ed.; Kluwer Academic Publishers: The Netherlands, 1992; pp 167-179.
6.   Overberg, A.; Hassenburger, A.; Hillenkamp, F.; In *Mass Spectrometry in the Biological Sciences; A Tutorial*; Gross, M.L., Ed.; Kluwer Academic Publishers: The Netherlands, 1992; pp 181-197.
7.   Campana, J.E., Ed.; In *Time-of-Flight Mass Spectrometry; Anal. Instrum.*, Marcel Dekker, Inc., New York, N.Y., 1987; Vol. 16, No. 1.
8.   Price, D.; Milnes, G.J.; *Int. J. Mass Spectrom. Ion Processes* **1990**, *99* 1-39.
9.   Van Vaeck, L.; Van Roy, W.; Gijbels, R.; Adams, F.; In *Laser Ionization Mass Analysis*; Vertes, A; Gijbels, R.; Adams, F., Eds.; Chemical Analysis; John Wiley and Sons: New York, NY, Vol. 124; pp 6-126.
10. *Lasers and Mass Spectrometry*; Lubman, D. M., Ed.; Oxford Series on Optical Sciences; Oxford University Press, Inc.: New York, New York, 1990.
11. Laude, Jr., D.A.; Hogan, J.D., *Technisches Messen*, **1990**, *57 (4)*, 155-159.
12. Brown, C.E.; Smith, M.J.C; *Spectros. World*, **1990**, *2 (1)*, 24-30.
13. Marshall, A.G.; Schweikhard, L; *Int. J. Mass Spectrom. Ion Processes*, **1992**, *118/119*, 37-70.
14. Koster, C.; Kahr, M.S.; Castoro, J.A.; Wilkins, C.L.; *Mass Spectrom. Rev.* **1992**, *11*, 495-512.
15. Annan, R.S.; Kochling, H.J.; Hill, J.A.; Biemann, K.; *Rapid Commun. Mass Spectrom.* **1992** *6*, 298-302.
16. Chambers, D.M; Goeringer, D.E.; McLuckey, S.A; Glish, G.L.; *Anal. Chem.* **1993**, *65*, 14-20.
17. Kebarle, P.; Tang, L.; *Anal. Chem.* **1993**, *65*, 972A-986A.
18. Beu, S.C.; Senko, M.W.; Quinn, J.P.; McLafferty, F.W.; *J. Am. Soc. Mass Spectrom.* **1993**, *4*, 190-192.
19. Nohmi, T.; Fenn, J.B.; *J. Am. Chem. Soc.* **1992** *114*, 3241-3246.
20. Smith, R.D.; Cheng, X.; Bruce, J.E.; Hofstadler, S.A.; Anderson, G.A. *Nature*, **1994** *369*, 137-139.
21. Simonsick Jr., W.J.; Prokai, L.; in T. Provder, M. Urban and H.G. Barth (Editors), Hyphenated Techniques in Polymer Characterization (American Chemical Society Symposium Series) American Chemical Society, Washington, DC, 1994.

22.   Smith, R.D.; Wahl, J.H.; Goodlett, D.R.; Hofstadler, S.A; *Anal. Chem.* **1993** *65*, 574A-584A.
23.   Nuwaysir, L.M.; Wilkins, C.L.; *Anal. Chem.* **1988**, *60*, 279-282.
24.   Busch, K.L.; Glish, G.L.; McLuckey, S.A. *Mass Spectrometry/Mass Spectrometry: Techniques and Applications of Tandem Mass Spectrometry;* VCH Publishers, Inc.: New York, NY, 1988.
25.   Brenna, J.T. In *Analytical Fourier Transform Ion Cyclotron Resonance Mass Spectrometry*; Asamoto, B.; VCH Publishers: New York, New York, 1992, pp. 187-213.
26.   Brenna, J. T. In *Microbeam Analysis-1989*; Russell, P.E., Ed.; San Francisco Press, Inc.: San Francisco, CA, 1989; pp. 306-310.
27.   Creasy, W. R.; Brenna, J. T. In *Polyimides: Mat., Chem., Charac.*; Feger, C.; Khojasteh, M. M.; McGrath, J. E., Eds.; Elsevier Science Publishers: B.V., Amsterdam, 1989; 635-642.
28.   Brown, C. E.; Kovacic, P.; Welch, K. J.; Cody, Jr., R. B.; Hein, R. E.; Kinsinger, J. A. *Polym.-Plast. Technol. Eng.* **1988**, *27*, 487-507.
29.   Brown, C. E.; Kovacic, P.; Welch, K. J.; Cody, Jr., R. B.; Hein, R. E.; Kinsinger, J. A. *J. Polym. Sci. Chem. Ed.* **1988**, *26*, 131-148.
30.   Brown, C. E.; Kovacic, P.; Wilkie, C.A.; Cody, R.B.; Hein, R.E.; Kinsinger, J. A. *Synth. Metals* **1986**, *15*, 265-279.
31.   Kahr, M.S.; Wilkins, C.L.; *J. Am. Soc. Mass Spectrom.* **1993**, *4*, 453-460.
32.   Brown, R. S.; Weil, D. A.; Wilkins, C. L. *Macromolecules* **1986**, *19*, 1255-1260.
33.   Brenna, J.T.; in *Microbeam Analysis--1989;* Russell, R.E., Ed.; San Francisco Press: San Francisco, CA, 1989; pp 306-310.
34.   Ijames, C. F.; Wilkins, C. L. *J. Amer. Chem. Soc.* **1988**, *110*, 2687-2688.
35.   Krier, G.; Pelletier, M.; Muller, J. F., Lazare, S.; Granier, V.; Lutgen, P. In *Microbeam Analysis--1989;* Russell, P.E., Ed.; San Francisco Press, Inc.: San Francisco, CA, 1989; pp. 347-349.
36.   Cody, R.B.; Kinsinger, J.A.; Ghaderi, S.; Amster, I.J.; McLafferty, F.W.; Brown, C.E.; *Anal. Chim. Acta* **1985** *178*, 43-66.
37.   Cromwell, E.F.; Reihs, K.; de Vries, M.S.; Ghaderi, S.; Wendt, H.R.; Hunziker, H.E.; *J.Phys. Chem.*, **1993**, *97*, 4720-4728.
38.   Brown, C. E.; Wilkie, C. A.; Smukalla, J.; Cody, Jr., R. B.; Kinsinger, J. A. *J. Polym. Sci., Poly. Chem.* **1986**, *24*, 1297-1311.
39.   Brown, C. E.; Kovacic, P.; Cody, Jr., R. B.; Hein, R. E.; Kinsinger, J. A. *J. Polym. Sci. Lett. Ed.* **1986**, *24*, 519-528.
40.   Brown, C. E.; Kovacic, P.; Wilkie, C. A.; Kinsinger, J. A.; Hein, R. E.; Yaniger, S. I.; Cody, Jr., R. B. *J. Polym. Sci. Chem. Ed.* **1986**, *24*, 255-267.
41.   Brown, C. E.; Kovacic, P.; Wilkie, C. A.; Cody, Jr., R. B.; Kinsinger, J. A. *J. Polym. Sci. Polym. Lett. Ed.* **1985**, *23*, 453-463.
42.   Ghaderi, S. *Ceramic Transactions* **1989**, *5*, 73-86.
43.   Nuwaysir, L. M.; Wilkins, C. L.; Simonsick, Jr., W. J. *J. Am. Soc. Mass. Spectrom.* **1990**, *1*, 66-71.
44.   Asamoto, B.; Young, J. R.; Citerin, R. J. *Anal. Chem.* **1990**, *62*, 61-70.
45.   Waddell, W. H.; Benzing, K. A.; Evans, L. R.; Mowdood, S. K.; Weil, D. A.; McMahon. J. M.; Cody, Jr. R. H.; Kinsinger, J. A. *Rubber Chem. Tech.* **1991**, *64*, 622-634.
46.   Hsu, A.T.; Marshall, A.G.; *Anal. Chem.* **1988**, *60*, 932-937.
47.   Johlman, C. L.; Wilkins, C. L.; Hogan, J. D.; Donovan, T. L.; Laude, Jr. D. A.; Youssefi, M. J. *Anal. Chem.* **1990**, *62*, 1167-1172.
48.   Xiang, X.; Dahlgren, J.; Enlow, W. P.; Marshall, A. G. *Anal. Chem.* **1992**, *64*, 2862-2865.

RECEIVED August 10, 1994

Hyphenated Thermal–Spectroscopic Techniques

Chapter 6

Polymer Characterization by an Advanced Simultaneous Thermogravimetric–Mass Spectrometric Skimmer Coupling System

E. Kaisersberger, E. Post, and J. Janoschek

NETZSCH-Gerätebau GmbH, P.O. Box 1460, D-95088 Selb, Germany

A top-loading simultaneous TG-DSC (STA) analyzer is combined with a quadrupole mass spectrometer (MS) by means of a two-stage gas inlet system. Evolved gases and vapours are collected just above the sample crucible by an orifice. The skimmer is arranged in the compression zone behind this orifice to achieve a parallel molecular beam into the electron impact ion source of the MS. The high sensitivity of this instrument combination is demonstrated by the ability to detect curing agents in rubbers and bromine flame retardants in electronic waste. The importance of a vacuum-tight construction and thus the purity of the selected sample atmosphere is shown by the thermal stability of a blend of polypropylene and melamine resin.

The well-known advantages of a quadrupole mass spectrometer (QMS) as a universal gas analysis instrument have often been used in combination with thermoanalytical methods. Especially interesting is the coupling of a mass spectrometry (MS) with thermogravimetry (TG,TGA) or simultaneous TG with Differential Scanning Calorimetry (DSC).

Most thermogravimetric experiments are carried out under atmospheric pressure in the sample chamber, whereby an exactly defined type of gas is often required. The working pressure of the QMS is approx. 10^{-5} mbar. Therefore a pressure reduction to the working pressure of the QMS must be realized in the coupling system which connects the thermal analyzer to the mass spectrometer.

The simplest way of coupling is by using a heated capillary with laminar flow conditions to extract the gas from the TG, and an orifice arranged at the outlet side of the capillary for a molecular flow into the ionization chamber of the MS. This coupling needs little preliminary modification of the thermoanalytical instrument. The capillary can be heated to 150°C or 200°C, but this cannot exclude condensation effects in some applications. The length of the capillary, usually 1m, can cause time or mass dependent separation of gases when they pass through it.

0097–6156/94/0581–0074$08.00/0

In order to avoid these disadvantages, NETZSCH and Balzers developed a practically perfect coupling system on the basis of two Al_2O_3 orifices for an application range of up to 1500°C *(1)*. This development took place 18 years ago. Based on long experience gained from countless applications, especially in high tech materials, a Skimmer coupling system was developed which can be used up to 800°C. The fields of polymer characterization, applied chemistry and environmental analysis place the main emphasis of skimmer coupling applications within this temperature range.

Experimental

The Skimmer coupling system *(2)* is based on the NETZSCH thermobalance STA 409 with top loading sample arrangement for TG or simultaneous TG-DSC.

The first step of the gas inlet system is arranged directly above the sample. Due to the vertical gas flow in the sample chamber and the short distance to the orifice, a total concentration of the evolved gases is guaranteed at this first pressure reduction step. The orifice is constructed as a divergent nozzle to optimize the flow conditions for the passing gas. The second pressure reduction step is formed by the Skimmer ("stream peeler") which is placed in the compression zone behind the nozzle (Figure 1). The dimensions of the vacuum devices and shape and arrangement of the pressure reduction steps result in the formation of a mainly parallel aligned molecular stream through the Skimmer.

The high mechanical precision, a prerequisite for the optimum performance of this type of coupling aiming in perfect gas flow conditions, could up to now only be achieved by using corrosion resistant stainless steel. Thus the Skimmer coupling can be applied up to 800°C in inert and oxidizing atmosphere. The fact that the coupling

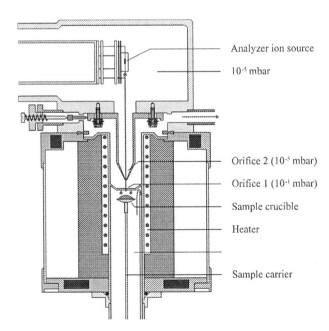

Analyzer ion source

10^{-5} mbar

Orifice 2 (10^{-5} mbar)

Orifice 1 (10^{-1} mbar)

Sample crucible

Heater

Sample carrier

Figure 1. Schematic drawing of the skimmer coupling system within the STA furnace.

system is extensively heated by the special furnace of the thermoanalyzer excludes condensation effects during gas transfer to the MS. The short distance between sample and ion source of the QMS avoids time dependent separation of the gases and provides a non-detectable time lag between gas evolution and detection in the MS. Together with the optimized gas flow conditions, an increased detection efficiency is achieved especially for condensable gases and vapours.

Results and Discussion

Thermogravimetry is used in the polymer field for the determination of the thermal stability, the quantitative compositional analysis of plastics and rubbers as well as of blends and for the detection of volatiles in vulcanized materials.

Figure 2 shows the TG and MS results on a peroxide-cured EPDM rubber. This experiment was conducted under pyrolytic conditions using a helium gas flow. The sample weight was 37.52 mg and the heating rate 10 K/min. The TG signal shows a two-step decomposition which is usually interpreted as the plasticizer evolution at the first step (here 240°C) and the decomposition of EPDM in the second step (here at 460°C). In this example, the combined techniques provide additional information. Scans were made during the heating at defined temperatures. The symbols "star" and "plus" stand for the intensity of the mass numbers 135 and 136 (atomic mass units, amu), respectively. The symbols are connected by lines as a guide to the eye. These mass numbers stand for cumyloxy-radicals from the dicumylperoxide (DCP) curing agent used for this rubber.

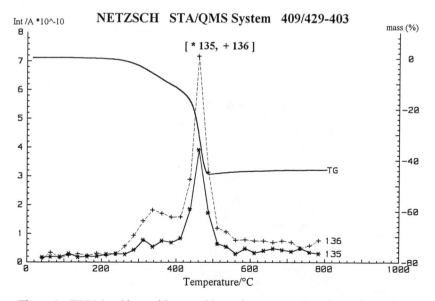

Figure 2. EPDM rubber with peroxide curing agent, detection of cumyloxy fragments.

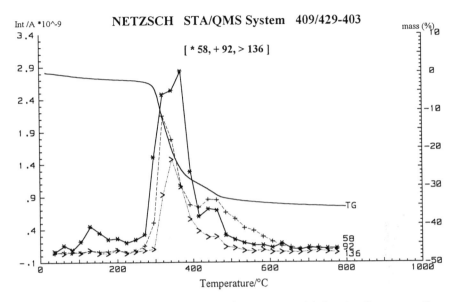

Figure 3. STA/MS results of an electronic waste material showing fragments of epoxy resin.

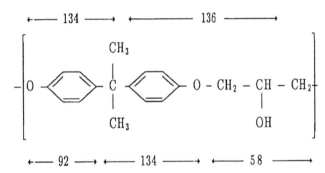

Figure 4. Example of a possible fragmentation of epoxy resin.

Without the additional QMS results, the first mass loss in the range 240°C to 400°C would only have been attributed to the content of plasticizer. In this case, the coupling of the techniques prevents, on the one hand, a misinterpretation of the results and, on the other hand, it allows an optimization of the curing process by adjusting the amount of DCP added to the rubber prior to vulcanization.

Electronic scrap is proving to be an ever growing problem for waste management. The materials contained in electronic waste are heterogeneous mixtures of various plastics, metals and other inorganic components (e.g. glass, ceramics...).

Only a very small part of the electronic waste can be reused. Most of it is disposed of thermally, e.g. for the reclaiming of noble metals. Many varied by-products which can prove to be dangerous for mankind and for the environment result from burning these heterogeneous mixtures.

Burning or pyrolytic procedures can be simulated by using the NETZSCH STA 409/QMS with the Skimmer coupling. In addition to detecting weight loss and

DSC or DTA signals, the corresponding gaseous decomposition products can be identified by mass spectrometry.

STA/MS results of an electronic waste sample from the automotive industry are shown in Figure 3. In this case, mostly circuit board material has been used. The sample was heated with 10°C/min in dynamic helium atmosphere. QMS data acquisition was carried out in SCAN mode. In 24°C intervals, the mass range of m/e = 6 to 200 was run through with a scanning rate of 0.1 sec/amu.

Three distinct weight loss steps are visible in the TG and DTG curves. The mass numbers shown represent fragments of epoxy resin (Figure 4).

The mass numbers 79, 80 and 82 are depicted in Figure 5. They show two peak maxima in the temperature range 250 ... 550°C. As the sample contains brominated epoxy resin (flame retardant function of bromine compounds), these fragments are $^{79}Br^+$, $H^{79}Br^+$ (m/e = 80) and $H^{81}Br^+$ (m/e = 82) as well as aromatic fragments.

The same sample was treated with a mixture of mineral acids in order to remove the metal parts from the batch. The STA/QMS measurements show distinct differences (Figure 6). In comparison to the untreated sample, a much greater weight loss is registered in the temperature range RT ... 200°C. In addition, the fragments 79, 80 and 82 are determined with QMS. This indicates that the thermal stability of the polymer is clearly reduced by the sample preparation with the mineral acid mixture. These are examples showing that the NETZSCH STA 409/QMS (skimmer coupling) can play an important part in the optimization of electronic waste management.

The relative thermal stability of the blend components in a blend of polypropylene (PP) and melamine resin is shown in Figure 7. The sample (8.82 mg) was heated in an inert atmosphere (N_2, 200 ml/min) to 800°C at a heating rate of 10 K/min. Significant weight loss starts at 240°C, showing two DTG peaks at 344°C and 453°C (when evaluated correspondingly) and a total mass loss of 99.74%. The first

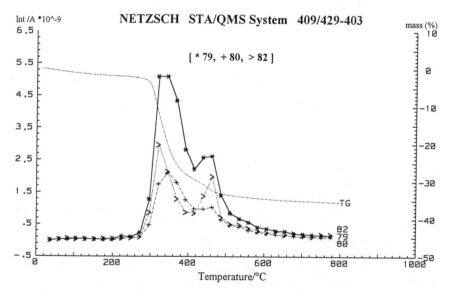

Figure 5. Untreated electronic waste material showing bromine fragments.

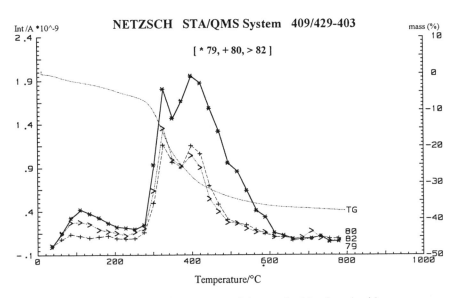

Figure 6. Electronic waste material treated with mineral acids.

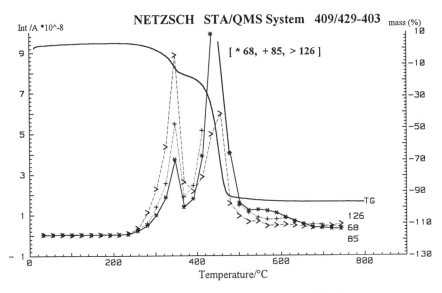

Figure 7. Pyrolysis of a PP/melamine resin blend.

TG step is 20.38%. The selected mass numbers 68, 85, 126 amu are found for PP as well as for melamine resin, but the relative height of the peaks at the two main TG steps indicates clearly that first melamine decomposes at 344°C (peak of the molecule ion 126 amu of melamine) and then PP is pyrolyzed (peak at 453°C, mainly mass numbers 68 and 85 amu, both overflow).

A series of separate plots for other fragments would prove this statement of a lower thermal stability of melamine in an inert atmosphere compared to PP.

Residual oxygen in the coupled TG-MS system would lower the decomposition range of the PP to the temperature of the melamine decomposition. This would exclude a separate evaluation of the melamine resin content (here 20.38%) in the blend. The vacuum-tight construction of the STA 409 Skimmer coupling system allows the precise atmosphere control and, in this example, an accurate determination of the relative thermal stability of the blend components.

Summary

The combination of thermoanalytical methods with quadrupole mass spectrometry results in an incredible improvement in the information received. Many application possibilities are of extreme topical relevance, such as the detection of nitrosamine precursory compounds during rubber vulcanization (originating from vulcanization agents), the determination of toxic or environmentally damaging exhaust gases during technical burning processes (polycyclic aromatic compounds, PCB, etc.) and in recycling projects. The optimization of both product and process can be counted as equally important in application ranges of basic materials and structural chemical research.

Literature Cited

(1)   Emmerich, W.-D.; Kaisersberger, E. *J. Thermal Anal.* **17**, *1979*, pp. 197.
(2)   Kaisersberger, E.; Emmerich, W.-D. *Thermochim. Acta* **85**, *1985*, pp. 275-278.

RECEIVED July 26, 1994

Chapter 7

Polymer and Other Degradation Studies Using Thermal Analysis Techniques

John P. Redfern[1] and Jay Powell[2,3]

[1]Rheometric Scientific (formerly Polymer Laboratories Ltd.) Thermal Sciences Division, Surrey Business Park, Kiln Lane, KT17 1JF Epsom, England
[2]Bio-Rad, Digilab Division, 237 Putnam Avenue, Cambridge, MA 02139

Polymer characterisation, stabilisation and degradation are very widely studied by Thermal Analysis (TA). Single techniques, such as thermogravimetric analysis (TG), differential scanning calorimetry (DSC), dynamic mechanical thermal analysis (DMTA), and dielectric thermal analysis (DETA) provide important information on the thermal behaviour of materials. However, to obtain a more complete profile of, say, polymer degradation gas analysis is required, particularly since all of the techniques listed give mainly physical information on the behaviour of materials.

The use of a simultaneous thermal analyser (STA) which is usually a combined simultaneous measurement of weight and energy usually referred to as a TG-DSC instrument) coupled to a mass spectrometer or to an FTIR provides a very powerful system for such studies. Key elements of the design will be stressed - underlining the importance of the interface system and the advantages of a comprehensive software package.

A number of specific studies are discussed in some detail and reference is made to other studies.

To say that plastics impacts all our lives is to state the obvious. The plastics industry is big business - a $150 - 180bn dollar business (equivalent to the Gross Domestic Product of a country like Switzerland) and growing worldwide both through increased usage in more and more countries and through new applications. The major end uses are shown in Table 1. There is also growing concern about disposal of used plastics and of the impact on the environment. There is, therefore, a clearly identified need for reliable methods to study the stabilisation and characterisation of these materials, to obtain knowledge on their properties and behaviour, the effects of modifying structure, additives and processes to produce the most appropriate, cost effective material for a specific requirement. The study of the degradation behaviour

[3]Corresponding author

0097–6156/94/0581–0081$08.36/0
© 1994 American Chemical Society

TABLE 1

Major End-Uses for Plastics
Packaging Construction Electrical and Electronic Industries Automotive Industry
Annual World Sales of $150-180 bn

TABLE 2a

Thermoplastics	Thermosets
Melting Crystallisation Glass Transition (T_g)	Curing Reactions
Expansion and Shrinkage Softening Strength Heat Capacity Thermal Conductivity Solvent and Water Retention	

TABLE 2b

Thermoplastics and Thermosets
Ageing Thermal Stability Thermal Degradation Chemical Degradation

is also important to contribute to our understanding of three possible decomposition scenarios, namely, in normal usage at operating temperatures, upon disposal or in accidental occurrences, for example in a fire. Therefore, there is a need to study the effects of ageing, the thermal stability, the degradation processes and the products of decomposition under a wide range of conditions.

The techniques of thermal analysis are very significant to the whole field of polymers in that they provide essential information relating both to their characterisation and their degradation. Table 2a lists areas where Thermal Analysis (TA) provides information for characterisation while Table 2b lists areas where TA provides information on the degradation processes.

TA techniques are a group of techniques in which the property of a sample is monitored against time or temperature while the temperature of the sample, in a specified atmosphere, is programmed. This programme may involve heating or cooling at a fixed rate of temperature change, or holding the temperature constant, or any sequence of these. The principal techniques are listed in Table 3. Of these the most useful in the study of the degradation process are **TG, DSC,** Simultaneous Thermal Analysis or **STA** (a combined simultaneous thermobalance with a DSC measuring head incorporated, abbreviated as **TG-DSC), STA-MS** (a TG-DSC instrument coupled with a mass spectrometer) and a **TG-FTIR** or **STA-FTIR** (a thermobalance or a STA linked with an FTIR spectrometer). This review is limited primarily with the application of STA, STA-MS and STA-FTIR to the degradation of polymers and other materials. STA provides a precise simultaneous understanding of the physical phenomena of weight and energy change. The STA coupled to a mass spectrometer or to an FTIR instrument gives a more complete profile of the degradation process.

TABLE 3

COMMON THERMAL ANALYSIS TECHNIQUES

TECHNIQUE	ABBREVIATION	MONITORS
SINGLE TECHNIQUES		
Thermogravimetry (Thermogravimetric Analysis)	TG	Mass
Differential Scanning Calorimetry	DSC	Energy
Differential Thermal Analysis	DTA	Temperature Difference
Thermomechanical Analysis	TMA	Dimensions
Dynamic Mechanical Thermal Analysis (Dynamic Thermal Analysis)	DMTA DMA	Response to Oscillatory Load
Dielectric Thermal Analysis	DETA	Response to Alternating Current
Evolved Gas Analysis	EGA	Nature and/or Amount of Volatiles
Temperature Programmed Reduction	TPR	Solid-Reducing Gas Interaction
SIMULTANEOUS TECHNIQUES		
Thermogravimetry-Differential Scanning Calorimetry	TG-DSC	Mass and Energy
+ Mass Spectrometry	TG-DSC-MS	Mass, Energy & Gas Analysis
+ FTIR Spectroscopy	TG-DSC-FTIR	Mass. Energy & Gas Identification

Instrumentation

A. Simultaneous Thermal Analyser (TG-DSC). The instrumentation for STA is shown schematically in Fig. 1. The advantage of single sample simultaneous experiments have been discussed previously (1, 2). It is appropriate to summarise these advantages:

1   Removes uncertainty - a) precise correlation of events occurring on both TG and DSC assured, b) removes any problems relating to sample inhomogeneity or batch variations.

2   Gives fuller characterisation - ensures greater certainty in identifying thermal events.

3   Validates quantitative measurements from DSC for phase changes, melting and purity measurements.

4   Detects moisture content enabling in-situ dry starting weight to be known accurately.

5   Accurate TG temperature calibration using DSC/DTA standard materials.

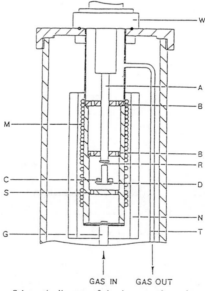

GAS IN   GAS OUT

Schematic diagram of simultaneous thermal analyser furnace and head. A, Ceramic tube to protect hangdown; B, movable baffle plates with gas ports; C, sample and reference crucibles; D, rigid heat flux TG–DSC plate; G, gas entry tube; M, mineral insulated graded heating element; N, liquid nitrogen cooling jacket; R, four bore ceramic hangdown suspended from electronic microbalance; S, fixed compartment divider with gas ports; T, side branch gas exit pipe; W, water-cooled cold finger.

Figure 1. Schematic diagram of a simultaneous thermal analyser. (Reproduced with Permission from Rheometric Scientific)

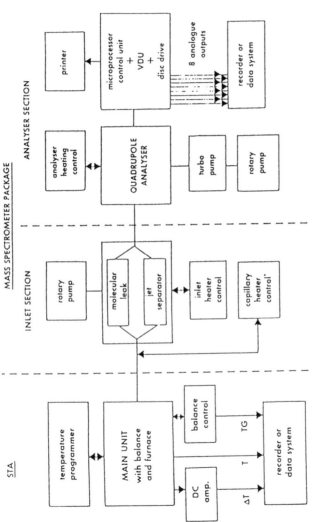

Figure 2. Schematic diagram of a STA-MS system (Reproduced with Permission from Rheometric Scientific)

B. Linked STA with Evolved Gas Analyser. For equipment to function in an optimum fashion it must be designed as a system. The design of the interface and the computing facility must be an integral part and must be appropriate to the instrumentation. The ideal requirements have been set out by Redfern (3) as follows:

1   Both instruments should operate under their respective optimum conditions - often greatly different.

2   The transfer time of the evolved species from the thermal analyser to the evolved gas analyser should be as low as possible.

3   The species to be analysed in the gas analyser should be in sufficient concentration, and mixing should be avoided for accurate detection and analysis.

4   To minimise possible secondary reactions between what maybe a considerable number of different evolved species by having a short transit time from thermal analyser to the spectrometer. Many reactions occurring on heating are much more complicated and give rise to more products than might be suggested from the curves obtained from the thermal analyser alone.

5   There is also the possibility of the capture of species by the walls of the interface - a particular problem with both moisture and with HCl, which also requires a short transit time and attention to the material of the transfer line.

6   To minimise the possibility of clogging of the interface by solid particles generated from the thermal events. The diameter of the interface needs to be chosen carefully so as to keep transfer time to a minimum (see 2 above) at the same time ensuring that smoke, etc., does not build up on the walls.

7   To ensure that control, data acquisition and processing are carried out by a single computer with an integrated software package, otherwise a lot of the advantages of the combined instrumentation will be lost.

A schematic of a STA-MS system is shown in Fig. 2 while that of a TG-FTIR system is shown in Figs 3 and 4. For more detailed information on the instrumentation see ref. 2. A number of applications of these types of instrumentation are listed in Ref. 3. Nagayama and Takada describe the direct combination of a thermobalance and an atomic absorption spectrometer for the detection of atomic vapour (4). Charsley et al. use a very simple capillary interface with a thermobalance and a quadrupole mass spectrometer to study both inorganic and polymer systems (5). In particular, they studied ethylene vinyl acetate (EVA) copolymers and poly(ethylene oxide). Systems have been proposed linking a thermobalance with a gas chromatograph (6). At first, this would appear to be difficult since gas chromatography is a discontinuous process. However, by using a fast column technique and an accurate sampling valve the possibilities are exciting.

Applications

A. General. i) The DSC and TG curves obtained from the appropriate thermal analyser often appear relatively simple. However, the thermally induced reaction may be much more complicated than the TA curve would suggest. Information from TG-DSC may not be able to explain the events fully. Thus, the use of a suitable gas

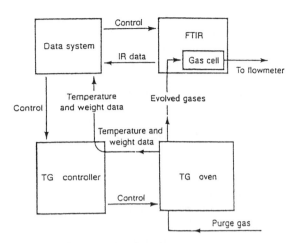

Figure 3. Schematic diagram of an integrated TG-FTIR system. (Reproduced from ref. 8. Copyright 1989 BioRad, Digilab Division)

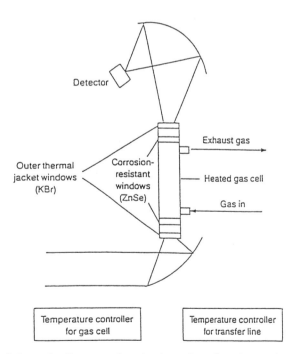

Figure 4. Schematic diagram of evolved gas interface bench for TG-FTIR system. (Reproduced by Permission from ref. 8. Copyright 1989 Biorad, Digilab Division)

analysis method is important to obtain a more complete understanding of the degradation behaviour of materials.

ii) The choice between a mass spectrometer and a FTIR spectrometer is determined by a number of factors including:

1   FTIR cannot detect gases which have no IR absorbance e.g. O_2, N_2. There can be problems with low absorbers e.g. H_2S. FTIR will not readily distinguish hydrocarbons above C_3H_8, since for the higher values the spectra coalesce.

2   MS detection levels are 2-3 orders of magnitude more sensitive than FTIR. MS requires special high vacuum capabilities and more stringent operating conditions.

3   Both MS and FTIR need the support of a full spectrum vapour phase library in order to maximise the information gained.

4   It follows that a mass spectrometer is particularly useful for studying the evolution of gases such as carbon dioxide, carbon monoxide, hydrogen and water from materials. Wherever non-IR absorbers are likely to be encountered, then MS is to be preferred.

5   In both cases the identification may be further enhanced by the use of a linked gas chromatograph as shown in Fig. 5. Either or both a and b paths may be followed. It should be recalled that GC is essentially discontinuous but separation of the gaseous components may be achieved by the use of a cold trap and subsequent analysis or by the use of a precision sampling system.

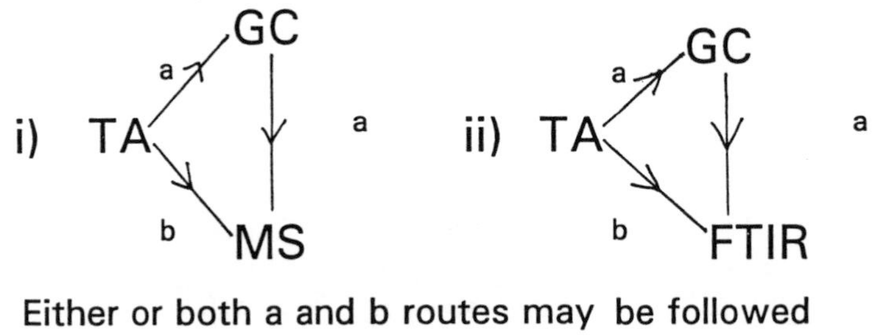

Either or both a and b routes may be followed

Figure 5. Multi-hyphenated techniques

6   As the strengths of MS and FTIR complement each other, then for complex degradation studies the linking of an STA to an FTIR spectrometer followed by a mass spectrometer will enable the most complete evaluations to be accomplished (7).

B. Results from TG-FTIR and STA-FTIR. It is possible to display a number of features of the decomposition pattern in real time and for each of these to be on the same time base for strict comparison purposes. Thus in addition to the weight profile (and its derivative) and the DSC signal, various gas evolution profiles can be plotted. The first of these is the Evolved Gas Profile (EGP) using the Gram-Schmidt orthogonalization method (8). This plots the mean of the integrated IR absorbance in the spectra as a function of time. It should be noted that the shape of the EGP curve is very similar to the derivative (DTG) curve although relative intensities may differ because each species evolved will have differing IR absorbances. Similarity on a time base is a <u>good</u> indication of no significant delay in passing from the thermal analyser to the FTIR via the heated transfer line. As many as five additional plots of up to 5 different IR regions may be displayed. Choosing the IR regions allows these plots to be specific to a particular molecule, thus the plot becomes a Specific Gas Profile (SGP). Alternatively, one window can display a narrow IR band profile - which is particular to a functional group i.e. Functional Group Profile (FGP). It is also apparent that for many of the evolved gases from TA experiments the gas may be identified not only by its characteristic frequencies but also by the band shape.

It is clear from the study of a number of materials by this technique (see ref. *3*) that an apparent single weight loss or DSC curve can conceal several different chemistries and the use of SGPs can allow for previously undetected subtleties in the reactions to be monitored.

The data obtained in this way may be further processed by the techniques of spectral search, spectral subtraction and spectral stripping. In spectral search; the spectra at a particular point in the decomposition is matched by comparison with a vapour phase library of spectra of a range of compounds. If there is a perfect match between the obtained spectra and a spectra contained in the library then it has a Hit Quality Index (HQI) of 0. Thus a HQI of 0 = perfect hit. A low value of the HQI of 0 - 0.3 combined with a substantial difference between the first and second hit (0.05 and upwards) indicates a very good match. If there is no compound match in the library then good functional group data may be obtained. By spectral subtraction the spectra of compounds which have been clearly identified are subtracted from the total spectral response using spectra recorded at the same gas cell temperature. This is done for the most commonly occurring gases in the combustion process e.g. CO, CO_2 and H_2O.

Using spectral subtraction techniques and the HQI approach to identify other compounds we can carry out successive subtraction techniques. This is called spectral stripping. By using a search algorithm that searches the derivative of the spectrum, rather than the original spectrum the computer tends to match the sharpest bands in the spectrum and gives less weight to the broader bands giving a more reliable identification.

B1. TG-FTIR. This can be illustrated by looking at the decomposition behaviour of polybutadiene in nitrogen (*8*) when it is heated at 20°C/min. The sample was heavily filled with inorganics resulting in a weight loss of 0.3 mg from the initial weight of 1.5 mg. The weight loss and derivative weight loss curve is shown in Fig. 6. The authors reported that the initial weight loss over the range 80 - 280°C is due to water loss together with some plasticizer. The higher weight loss from 400°C evolved a mixture of gases. The reported spectrum obtained at 500°C (see Fig. 7) includes bands which can be readily identified as due to carbon dioxide 2400 - 2280 cm^{-1}, carbon monoxide 2220 - 2060 cm^{-1} and water corresponding to the complex of sharp bands between 4000 and 3500 cm^{-1} and between 2000 - 1350 cm^{-1} which are all to be expected. There is a strong absorption at about 2900 cm^{-1} which shows the presence of hydrocarbons. The computer programme was used, to subtract the spectral bands of CO_2, CO and H_2O using the reference spectra recorded at the

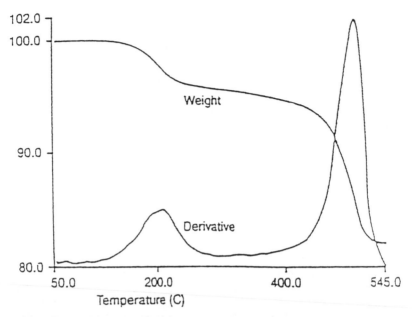

Figure 6. TG-FTIR of polybutadiene 1. TG and DTG curves of 1.5 mg at 20°C/min in flowing nitrogen (Reproduced by Permission from ref. 8. Copyright 1989 Biorad, Digilab Division)

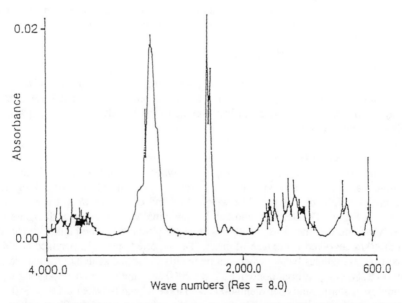

Figure 7. TG-FTIR of polybutadiene 2. IR spectrum from decomposition products at 500°C. (Reproduced by Permission from ref. 8. Copyright 1989 Biorad, Digilab Division)

same gas cell temperature. The resulting difference spectra was searched against the reference vapour phase spectral library using the derivative of the spectrum. This led to the identification of methane with its sharp band at 3010 cm⁻¹. The library spectrum of methane was then subtracted and the new difference spectrum was searched. This time the best hit was n-pentane followed in terms of HQI by n-butane and other similar hydrocarbons. The spectrum of n-butane as being the most likely product was now subtracted. The next search identified ethylene by its sharp band at about 950 cm⁻¹. This residual spectrum shows the presence of weak bands that lead to HQIs that are indicative of the presence of cyclic hydrocarbons. The procedure and the results obtained indicate that a single weight loss may well correspond to a very complex mixture of evolved gases. If the authors had used a slower heating rate more light might have been thrown on the complex reactions occurring.

Synthetic rubbers used by the automotive industry are somewhat similar to the polybutadiene discussed above. However as rubber formulations in the tyre industry become more complicated, the analysis of the finished product becomes more difficult. A 4 mg sample was heated from 100 - 600°C at 20°C/min in nitrogen. Total weight loss was approximately 66%; therefore, about 2.6 mg of gases were evolved from the decomposition of the sample. The derivative curve indicated two weight loss regions (Fig. 8). Spectra were selected across these two regions and were searched after subtracting the carbon dioxide spectra. The first

Weight loss for rubber sample

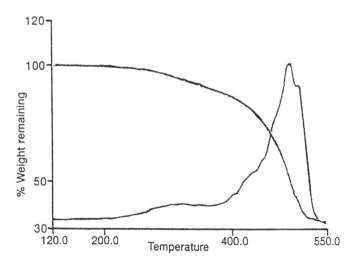

Figure 8. TG-FTIR of rubber 1. TG and DTG curves (Reproduced by Permission from ref. 8. Copyright 1989 Biorad, Digilab Division)

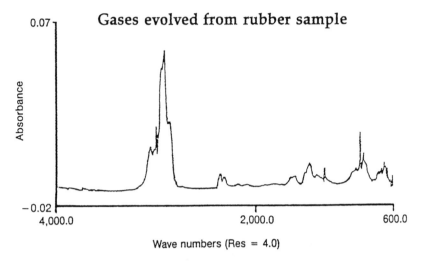

Figure 9. TG-FTIR of rubber 2. Gases evolved. (Reproduced by Permission from ref. 8. Copyright 1989 Biorad, Digilab Division)

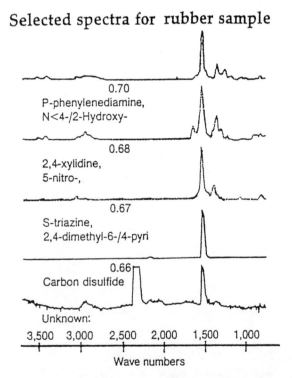

Figure 10. TG-FTIR of rubber 3. Selected spectra. (Reproduced by Permission from ref. 8. Copyright 1989 Biorad, Digilab Division)

region at around 250°C resulted from the evolution of carbon dioxide, carbon monoxide and carbon disulphide. There also appears to be a trace of hydrocarbon present. The major weight loss occurs between 375 and 540°C. This weight loss appears to be caused by the evolution of a short-chained alkene and methane. The loss of the alkene begins at 375°C and continues until the weight loss is complete. The methane evolution does not begin until approximately 450°C. The results are shown in Figs. 9 & 10.

B2. STA-FTIR. The above application study was carried out using an integrated TG-FTIR system. However, STA-FTIR studies have also been carried out using an STA (Polymer Laboratories 625) linked through a silica-lined stainless steel heated transfer line to a Digilab FTS spectrometer with attached TG interface bench. The interface contained a heated gas cell equipped with a, room temperature deuterated triglycine sulphate (DTGS) detector. Controllers for both the heated gas cell and transfer line maintained their temperatures at 230 and 220°C respectively (*9*). A 5.7 mg sample of zinc stearate was heated to 600°C in dry nitrogen at 20°C/min. The results are shown in Figs. 11 - 13. The DSC curve (Fig. 11) gave a sharp endotherm at 128°C corresponding to the melting point of the material. No weight loss was observed at this temperature, but the infrared spectrum recorded at the same time showed that a very small amount of water was released. Fig. 12 shows the 1700 - 1800 cm^{-1} region of the spectrum, integrated as a function of time. This region is where the expected stearic acid carbonyl spectrum should appear, but it is also sensitive to water vapour. A very small peak can be seen at 4.5 minutes, showing the release of water for less than a minute. Although the TG curve does not pick this up it should be remembered that the FTIR is sensitive to fractions of a microgram, whereas the STA is sensitive to 1μg.

Decomposition of this material commences below 300°C and takes place over a long temperature interval. The derivative shows two peaks rather than the one expected and is similar to the infrared profile shown in Fig. 12, except that it shows no peak corresponding to the water loss just mentioned. The infrared spectra of the two major weight losses are shown in Fig. 13. The two spectra are plotted on the same absorbance scale to show the relative amounts being evolved. The C-H stretching region (2800 - 3000 cm^{-1}) shows the band shape typical of a long hydrocarbon chain such as a stearate. Of particular interest is the carbonyl region of the spectrum (1700 - 1800 cm^{-1}). The spectrum recorded at 380°C shows a carbonyl maximum at 1775 cm^{-1} (see Fig. 13a), whereas the spectrum recorded at 440°C shows a second, stronger peak at 1717 cm^{-1} (see Fig. 13b). In addition the authors showed that both spectra displayed a weak peak at 3575 cm^{-1} that is due to the O-H stretching motion of stearic acid in a non-hydrogen bonded environment and where there is no water present in the system. Why is this? The expected product, stearic acid would be expected to show a band at about 1780 cm^{-1} but not at 1717 cm^{-1}. However an independent TG study (*10*) suggests that stearic acid is lost as an aerosol. So the second sharp band at 1717 cm^{-1} has been assigned to stearic acid in the condensed phase. The observation of the two carbonyl frequencies underlines that the gas transfer from the STA to the FTIR is excellent, since the stearic acid is observed in both gaseous and aerosol forms. Any flow restrictions or cold spots would be expected to give rise to a reduction in, or complete loss of, the amount of

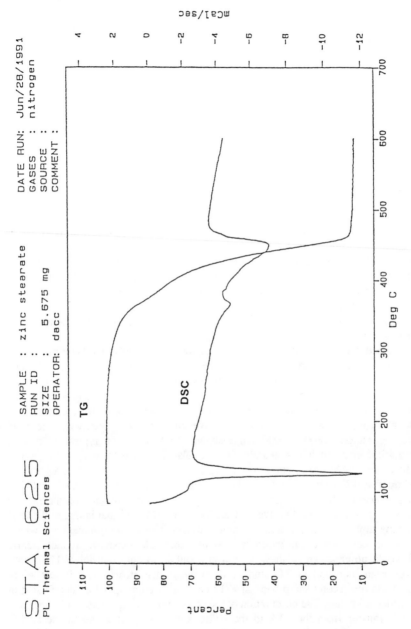

Figure 11. STA-FTIR of zinc stearate 1. TG and DSC curves

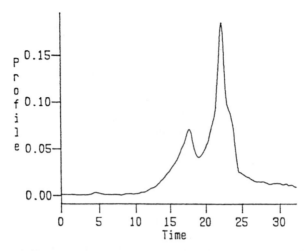

Figure 12. STA-FTIR of zinc stearate 2. Integrated spectrum over the 1700 - 1800 cm⁻¹ region

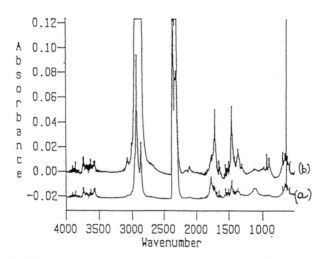

Figure 13. STA-FTIR of zinc stearate 3. Spectra taken at 380°C (curve a) and 440°C (curve b)

aerosol being delivered to the gas cell. Other minor components can also be identified from the spectra, presumably caused by the decomposition of the stearic acid at the high temperature. There is a single sharp band at 2120 cm⁻¹, which is possibly due to the presence of zinc carbonyl.

The above examples show the powerful potential of TG-FTIR and the STA-FTIR as analytical techniques and reinforce the view that the evolved gases are often much more complicated than would be expected from the TA curves alone. Many other materials have been studied (*8, 11, 12*) which show the wide applicability of this method.

C. Results from STA-MS. The mass spectrometer may be operated in one of two modes, either in the scanning mode or in the peak select mode. The first stage in the examination of the decomposition of a material would be to perform the STA experiment while recording sequential mass scans during any observed weight loss stages. Scan speeds are typically of a few seconds for the entire amu range of the mass spectrometer. Thus, many spectra can be recorded during a weight loss process. From an examination of these spectra it is usually possible to identify certain peaks as being significant. A second experiment is then conducted with the mass spectrometer in the peak select mode; in which a number of peaks (up to 16) may be continuously monitored as a function of time.

An example of the type of information generated by the STA-MS system is provided by a study of ripidolite, which is an iron-rich chlorite. The structure consists of alternate layers of mica-like sheets and metal cation-hydroxyl sheets with Fe(II) and Fe(III) associated with the hydroxyl groups of both layers. The weight loss from 470-750°C evolves water from the hydroxyl sheets and from 750-830°C the water is evolved from the mica sheets. The run was carried out in an inert atmosphere of nitrogen. The Fe(II), exposed during and after the first water loss, reduces some of the water evolved in the second stage to hydrogen. A continuous monitoring of mass numbers 2 and 18 shows this very clearly (Fig. 14).

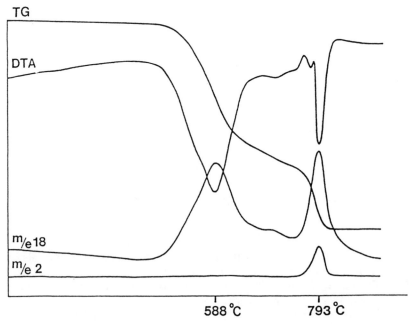

Figure 14. STA-MS of 46.5 mg of ripidolite at 15°C/min in flowing nitrogen.

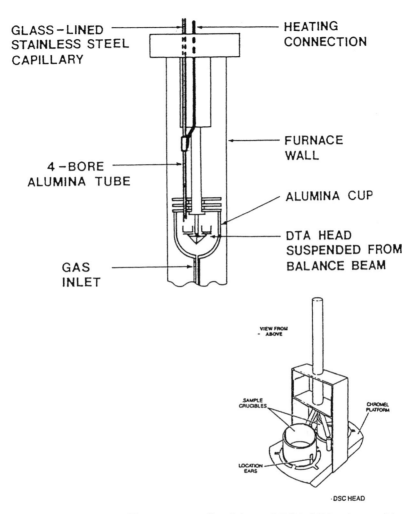

Figure 15. Heated capillary gas sampling inlet and TG-DSC head assembly (Reproduced with permission by Rheometric Scientific)

A further application is the use of this technique to study the pyrolysis of the polysilane ceramic precursor (*13*). Since the mid 1970's there have been great efforts to produce ceramics using polymeric metallo-organic precursors as starting materials. There are many problems using these novel routes, but great gains can be achieved e.g. higher material starting purity, lower costs in sintering final product etc. TA-MS has been used to examine the fundamentals of the ceramic forming polymer pyrolysis reaction in order to increase the yield of ceramic material derived by this route. The system used in the study was a Polymer Laboratories STA 1500HF linked to a Fisons Thermolab bench top 300 amu quadrupole mass spectrometer via a heated capillary interface line (see Fig. 3) The heated capillary (250°C) takes the

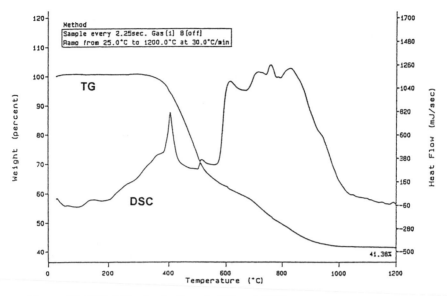

Figure 16. STA-MS of polysilane 1. TG and DSC curves

evolved gases directly from above the sample (Fig. 15) through to the heated molecular leak and then direct into the quadrupole MS. The transfer time was better than 100 milliseconds. A gas flow of 50ml/min dry argon was used with a heating rate the same as in commercial production, namely 30°C/min. Figure 16 shows the STA results. There appears to be a small weight loss around 150 - 200°C. The main weight loss commences about 380°C. There is then steady weight loss until about 500°C. Two further slower weight losses continue up until around 1000°C. The residue (41.38%) is a fine fibre material which was presumed to be SiC, although it was not characterised. The DSC curve shows two exothermic reactions. The first at around 400°C is very sharp, this is followed by a large, broad complex peak lasting up to the end of the reaction. Running the MS in the scan mode enables one to identify the major species. Figure 17 shows selected scan histograms at increasing temperatures throughout the run. Certain higher mass fragments appear at the higher temperatures. Certain specific masses were selected as a result of these preliminary scan runs. These were then monitored throughout the experiment on the second run. The results are shown in Fig. 18. The species evolved around 150°C were deduced to be side chain cleavage from the polysilane structure. This cleavage is also associated with the weight loss due to the evaporation of some low molecular weight compounds, mainly benzene (M/Z 78), toluene (M/Z 91) and some hydrocarbons. During the main reaction from 350°C upwards the most likely reaction is generally free radical in nature (14). Under these conditions the Si-C bond is relatively stable towards homolytic fission whereas the C-H bond in Si-CH$_3$ is readily broken. Thus the primary species are assumed to have M/Zs of 198, 183 and 168 respectively. It is interesting to note that the higher molecuar weight species only appear during the second stage reaction process. From the recombination of the radicals produced,

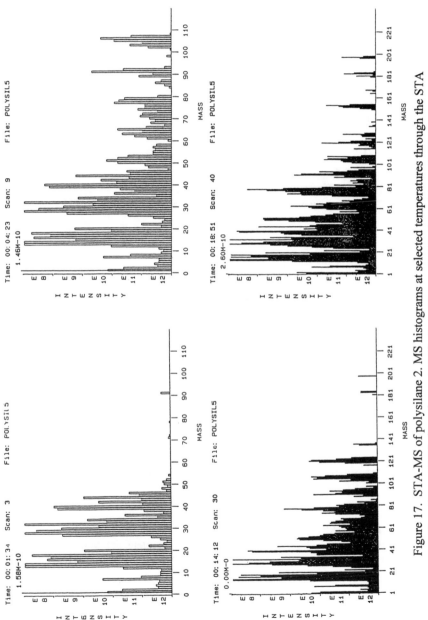

Figure 17. STA-MS of polysilane 2. MS histograms at selected temperatures through the STA run. Scan 3 = ~ 70°C, Scan 9 = ~ 150°C, Scan 30 = ~ 420°C, Scan 40 = ~ 620°C.

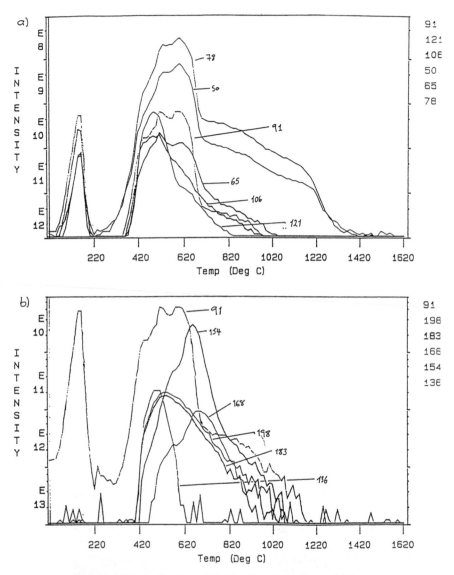

Figure 18. STA-MS of polysilane 3. MS selected species at a) lower molecular weights b) higher molecular weights. It is interesting to note that the higher molecular weight species only appear during the 2nd stage reaction process.

networks and three dimensional structures build up with the loss of methane and hydrogen as can be seen in the histograms in Fig. 17. The results give a valuable insight into some of the processes occurring in a far from understood material.

Conclusion

It is clear from the work reviewed here and from the additional references that an increasing range of materials are being studied by hyphenated TA techniques. The use of these techniques contribute greatly to understanding the effect of heating materials, whether this occurs in the course of normal use or whether it arises from accidental heating

Acknowledgments

The authors acknowledge the collaboration between Polymer Laboratories, Thermal Sciences Division (now Rheometric Scientific) and Fisons (the STA-MS System) and with BioRad (the TGA and STA-FTIR systems). In particular the help of David Compton, Paul Nicholas and Paul Larcey is gratefully acknowledged.

Literature Cited

1   John Redfern. *International Labmate*, **1986,** *11*(1), 19. Advantages and Applications of Simultaneous TG-DSC.

2   Jenny Hider. *International Labmate*, **1990,** *15*(6), . The Advantages of Simultaneous Thermal Analysis and the Technology behind it.

3   John P.Redfern. *Polymer International*, **1991,** *26*, 51-58. Polymer Studies by Simultaneous Thermal Analysis Techniques.

4   Kazuyuki Nagayama and Takeo Takada. *Thermochimica Acta* **1989,** *156*, 11-19. Direct combination of Thermogravimetric Analyzer and Atomic Absorption Spectrometer for Detection of Atomic Vapour in Thermal Analysis.

5   E.L. Charsley, S.B. Warrington, G.K.Jones and A.R. McGhie. *International Laboratory* **1990** March/April 13-19. Thermogravimetry - Mass Spectrometry using a simple Capillary Interface.

6   M. Kaljurand and E. Kullik. Computerised Multiple Input Chromatography. Ellis Horwood, **1989,** 225pp. (see esp. pp 162-185). .

7   A.S Mogaddan, Univ of Tarbiat-Modarres, Iran **1992.** Private Communication.

8   David A.C. Compton, David J. Johnson and Martin L. Mittleman. Bio-Rad FTS/IR Note No. 70 1989 (Reprinted from Research and Development Feb & April 1989). Integrated TGA-FTIR System to Study Polymeric Materials.

9   David A.C. Compton and David Loeb. 20th NATAS Conference **1991**. A combined TGA-DSC-FTIR System.

10   J.A.J. Jansen, J.H. van der Maas and A. Posthuma de Boer, **1991**. Private Communication,

11   D.J. Johnson and D.A.C. Compton. *American Laboratory* **1991**, (1) 37-43. Quantitative Analysis of Nitrocellulose and Pulp in Gunpowder using TGA-FTIR.

12   David J. Johnson and David A.C. Compton. *Spectroscopy* **1988,** *3*(6) 47. Solvent Retention Studies for Pharmaceutical Samples using an Integrated TGA-FTIR System.

13 P. Larcey. 21st NATAS Conference **1992**. The Use of Hyphenated TA-MS in the Study of the Pyrolysis of Polysilane Ceramic Precursor.
14 R. Bolton. Organic Mechanisms. Penguin Physical Science Text, London, **1972**.

RECEIVED August 22, 1994

Chapter 8

Decomposition of Ethylene–Vinyl Acetate Copolymers Examined by Combined Thermogravimetry, Gas Chromatography, and Infrared Spectroscopy

Brian J. McGrattan

Perkin-Elmer Corporation, 761 Main Avenue,
Norwalk, CT 06859–0240

By combining information from thermogravimetry/infrared spectroscopy and gas chromatography/infrared spectroscopy measurements of the trapped gases, characterization of complex polymer decompositions can be made. In this paper, the thermally induced breakdown of ethylene vinyl acetate (EVA) copolymers was studied. Results show that EVA undergoes a two-step decomposition: an acetate pyrolysis of the copolymer leaving a polyunsaturated linear hydrocarbon, followed by the breakdown of the hydrocarbon backbone to produce a large number of straight-chain hydrocarbon products.

Infrared spectroscopy and thermogravimetry have been used in polymer analysis for many years. By coupling the effluent of thermogravimetry to an infrared gas cell, TG/IR (sometimes known as evolved gas analysis) has been used to examine the thermally induced decomposition products a variety of polymers including of poly(vinyl chloride) (1), polyacrylamide (2), tetrafluoroethylene-propylene (3) and ethylene-vinyl acetate (4) copolymers, as well as styrene-butadiene composite (5). In most cases, the thermal decompositions lead to a coevolution of several materials, particularly when the backbone of the polymer is broken apart. Because of the mixture of products generated, identification of the constituents could only be made through the use of spectral subtraction or when small molecules with simple infrared spectra, such as carbon monoxide, carbon dioxide, and hydrogen chloride were generated. In the study on polyacrylamide (2), the effluent was trapped in a gas collection tube immersed in liquid nitrogen and then separated and analyzed by GC/FT-IR and gas chromatography/mass spectrometry.

In this paper, the thermally induced decomposition of ethylene-vinyl acetate copolymers is examined. The analysis of the trapped effluents by GC/FT-IR is significantly different than that reported by other authors (4). These authors noted the formation of acetic acid and a polyolefin at 360°C to 450°C, as is observed in

0097–6156/94/0581–0103$08.00/0

this paper. Subsequent decomposition of the polyolefin from 450°C to 550°C reportedly lead to 1-butene, ethylene, methane and carbon dioxide. Here, acetic acid, carbon dioxide and carbon monoxide are initially observed, followed by backbone fragments in the range of C_8 to C_{26}. These backbone fragments consist of a mixture of terminal dienes, terminal alkenes and alkanes. The distribution of the hydrocarbon fragments furthermore appears to be qualitatively indicative of the amount of vinyl acetate and the distribution of vinyl acetate in the copolymer.

Experimental

Samples of ethylene-vinyl acetate copolymers of nominal 12, 18, 25, 28, 33 and 40 weight percent vinyl acetate were obtained from Scientific Polymer Products, Inc. (Ontario, NY).

The thermal analysis system used was a Perkin-Elmer TGA7 (Norwalk, CT). Samples ranging from 20 to 40 milligrams of the samples were heated at a rate of 20°C per minute in a nitrogen atmosphere. The flow rate of nitrogen into the cell for the TG/IR experiments was approximately 80 milliliters per minute.

A trap was constructed by placing a glass wool plug, approximately 0.2 milliliters of Tenax, and another glass wool plug in a split capillary injector liner (4 millimeters internal diameter). The trap was fixed to the vent of the TG/IR accessory for collection of the sample, and desorbed by placing it in the capillary injector of the gas chromatograph and heating the injector ballistically to 250°C.

A Perkin-Elmer System 2000 FT-IR was fit with a TG/IR accessory and a GC/IR accessory. The TG/IR accessory used a deuterated triglycine sulfate (DTGS) detector. The spectra of the evolved gases were acquired and four scans co-added to generate each TG/IR time slice, corresponding to approximately a 3°C rise in TG program temperature. The GC/IR accessory was fit with a liquid nitrogen cooled, narrow band mercury-cadmium-telluride (MCT). GC/IR spectra were acquired, co-added and stored to disk at a rate of approximately 0.7 spectra per second.

A Perkin-Elmer AutoSystem Gas Chromatograph with a 25-meter long, 0.32-millimeter diameter column and 0.5-micron DB-1 film was used. At the onset of the thermal desorption of the Tenax trap, a temperature program of 35°C to 250°C at 5°C per minute was run.

The TGA-7 was used with the Perkin-Elmer PC Series multitasking software. The System 2000 FT-IR used the Time Resolved Infrared software for data acquisition of the GC/IR and TG/IR data, and the Quant-C curve-fitting quantitative analysis software. Perkin-Elmer PC Search software and the Sadtler Vapor Phase Library were used.

Results and Discussion

The TG weight loss curve and the Gram-Schmidt infrared thermogram are seen in Figure 1. Ethylene-vinyl acetate copolymers show two distinct weight losses with maximum rates at 364°C and 480°C. The spectra obtained at the maximum of the Gram-Schmidt thermogram for these two events are shown in Figure 2. The upper

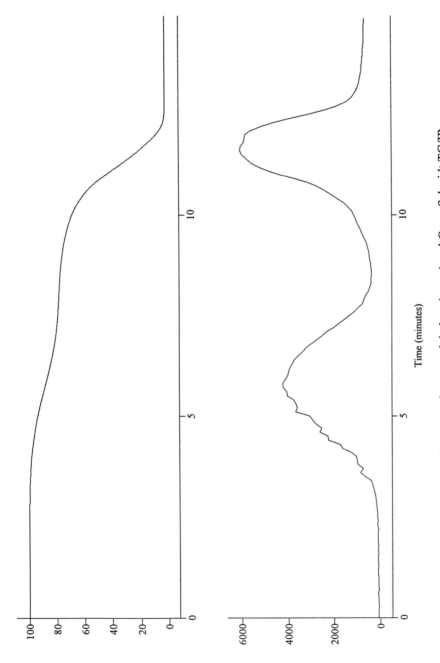

Figure 1. The thermogravimetry weight loss (upper) and Gram-Schmidt TG/IR (lower) curves for ethylene 33% vinyl acetate.

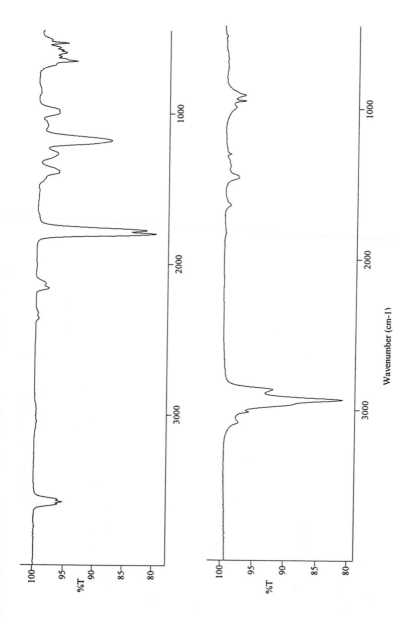

Figure 2.   IR spectra obtained at the maximum evolution rate for the two thermal events.   The upper spectrum is that of acetic acid with a small amount of carbon monoxide and carbon dioxide.   The lower spectrum is closely resembles that of 1-octene.

spectrum is primarily that of acetic acid, although small amounts of carbon monoxide (peaks at 2165 and 2132 cm^{-1}) and carbon dioxide (peaks at 2364 and 2330 cm^{-1}) are evident. Acetate pyrolysis yielding an alkene and acetic acid is a well known reaction (*6*). The CO and CO_2 evolution may be indicative of an alternate decomposition mechanism, or the combustion of materials from residual oxygen in the TGA system.

Table I shows the measured vinyl acetate content of each of the samples based on the stoichiometry of every vinyl acetate group decomposing to acetic acid and a polyolefin in the initial weight loss. The DY values are those measured as the changes in weight of the sample by TGA. With the exception of the 12% nominal vinyl acetate sample, relatively close agreement was found between the nominal values and the TGA data.

Table I. Nominal and measured concentrations of vinyl acetate in EVA copolymers.

Nominal vinyl acetate (Weight %)	ΔY Acetic Acid (Weight %)	Vinyl acetate (Weight %)
12	12.46	17.86
18	13.02	18.65
25	16.62	23.81
28	20.32	29.12
33	21.02	30.13
40	29.02	41.48

The second weight loss of EVA, occurring over the temperature range of 430°C to 510°C, produces the spectrum depicted in the lower portion of Figure 2 at the maximum weight loss rate. The spectrum appears to be that of a terminal alkene. Spectral library searching of the Sadtler Vapor Phase collection reported a best match of the spectrum with 1-octene.

More careful examination of the TG/IR data shows that the ratio of olefinic - CH to -CH_2- is changing during the secondary evolution. The spectra from 450°C to 490°C in 10°C intervals are shown in Figure 3. The intensity of the band at 3015 cm^{-1} (=CH_2 stretch) relative to the band at 2934 cm^{-1} (-CH_2- stretch) increases with increasing temperature, indicating that a mixture of materials are evolved. Three possibilities exist: if the compounds are all 1-alkenes, then the length of the chain decreases with increasing temperature; if the compounds are a mixture of alkanes and alkenes, then proportionally more alkane is evolved earlier in the run than later in the experiment or both the average chain length and the alkene to alkane ratio are changing with temperature.

In order to ascertain the composition of the mixture of hydrocarbons, the effluent of the TG/IR was captured on a trap constructed from a GC capillary injector liner with Tenax solid phase adsorbent and analyzed by GC/IR. The trapped

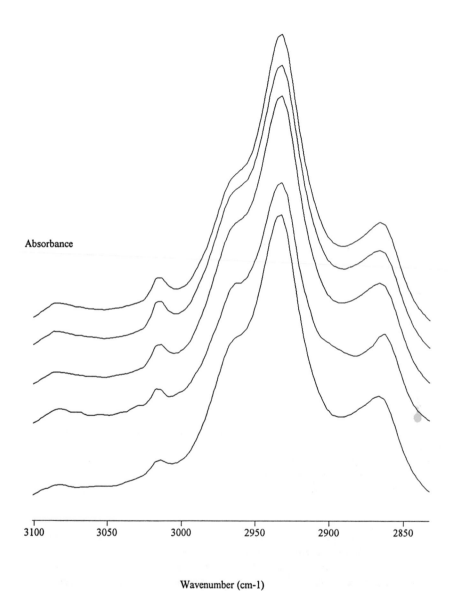

Absorbance

3100   3050   3000   2950   2900   2850

Wavenumber (cm-1)

Figure 3. CH stretch region of TG/IR time slices obtained at 450°C through 490°C (bottom to top) in 10°C intervals. Note the changing amount of $=CH_2$ at 3015 cm^{-1}.

Table II. Assignments of identities of chromatographic peaks.

Retention time (minutes)	Identity	Retention time (minutes)	Identity
0.84	carbon dioxide and water	30.80	octadecane
2.50	acetic acid	32.46	1,18-nonadecdiene
4.62	1-octene and octane	32.80	1-nonadecene
6.78	1-nonene and nonane	32.90	nonadecane
9.32	1-decene	34.50	1,19-eicosdiene
9.67	decane	34.80	1-eicosene
12.25	1-undecene	34.90	eicosane
12.56	undecane	36.42	1,20-heneicosdiene
15.14	1-dodecene	36.70	1-heneicosene
15.47	dodecane	36.83	heneicosane
17.99	1-tridecene	38.29	1,21-docosdiene
18.28	tridecane	38.50	1-docosene
20.69	1-tetradecene	38.61	docosane
20.86	tetradecane	40.06	1,22-tricosdiene
23.28	1-pentadecene	40.25	1-tricosene
23.58	pentadecane	40.38	tricosane
25.52	1,15-hexadecdiene	41.74	1,23-tetracosdiene
26.90	1-hexadecene	41.94	1-tetracosene
26.11	hexadecane	42.06	tetracosane
27.95	1,16-heptadecdiene	43.42	1,24-pentacosdiene
28.30	1-heptadecene	43.60	1-pentacosene
28.50	heptadecane	43.74	pentacosane
30.28	1,17-octadecdiene	45.54	1-hexcosene
30.60	1-octadecene	45.71	hexacosane

materials were desorbed directly onto a GC column by placing the injector liner into the GC injector and ballistically heating the injector to 250°C. Figure 4 shows the Gram-Schmidt chromatogram for one of the samples, the ethylene 33% vinyl acetate copolymer. Approximately 50 species were separated from the mixture of evolved gases. Peaks eluting in families of three were observed throughout the chromatogram, as seen in Figure 5. These families consisted of a terminal diene, a 1-alkene and an alkane. Chain lengths from C_8 to C_{26} were observed as determined by computation of the $=CH_2$ to $-CH_2-$ ratio for the alkenes and $-CH_3$ to $-CH_2-$ ratio for the alkanes, and by spectral library searching. The assignments, listed in Table II, were confirmed by injecting known materials into the GC and comparing spectra and retention times, and by GC/MS. It is possible that chains shorter than C_8 were evolved, but not trapped. Similarly, chains longer than C_{26} may have been evolved and lost prior to trapping, not thermally desorbed from the Tenax, or not eluted over the observed GC/IR acquisition time.

As mentioned earlier, the TG/IR time slices indicate that more olefinic =C-H stretching was observed as the temperature increased. A curve-fit of the GC/IR spectra of acetic acid, dodecane and dodecene against the TG/IR time slices was performed to determine the relationship in more detail. The results of this experiment are shown in Figure 6. As would be expected, the concentration of acetic acid increases early in the experiment, but by 400 °C, has decreased and returned to baseline levels. The curve fit for the two hydrocarbons shows that saturated hydrocarbons, as represented by dodecane, evolve at lower temperatures than the corresponding unsaturated hydrocarbons. Unfortunately because the choice of hydrocarbon standards assumes constant chain length, this observation does not suggest whether only proportionally more alkene is evolved later, or whether both the average chain length and the alkene/alkane ratio are changing. Although the two overlap in the Gram-Schmidt thermogram, the TG/IR time slices for the components can be resolved in this manner.

The trapping TG/GC/IR experiment was repeated for each of the EVA copolymers. Figure 7 shows the Gram-Schmidt chromatograms for the nominal 12% EVA and the 40% EVA. Qualitatively, the distribution of hydrocarbons for samples containing higher amounts of vinyl acetate shifts toward smaller fragments and vice versa. Since the retention times and number of methylene groups are related, computing the 25%, 50% and 75% total absorbance area from an infrared absorbance chromatogram for each sample should indicate the distribution of hydrocarbons. These data (Table III) show that the distribution is not very significantly different: for all of the copolymers, half of the hydrocarbon fragments were from C_{14} to C_{21}.

Table III. Distribution of hydrocarbons in trapped evolved gases.

Vinyl acetate (Wt %)	Retention 25% of total area	Carbon chain length	Retention 50% of total area	Carbon chain length	Retention 75% of total area	Carbon chain length
17.86	26.2	16.1	30.9	18.1	35.1	20.1
18.65	23.5	15.0	28.2	17.0	36.6	20.9
23.81	25.1	15.0	32.2	18.7	36.6	20.9
29.12	30.5	17.0	32.8	18.9	36.2	20.7
30.13	29.8	17.6	32.8	18.9	36.2	20.7
41.48	21.1	14.1	27.0	16.4	34.4	19.8

Conclusions

The utility of the trapping TG/GC/IR experiment in the characterization of complex polymer decompositions has been demonstrated. As shown in an earlier study, the initial decomposition follows an acetate pyrolysis elimination, producing acetic acid. Unlike the earlier study, however, the secondary decomposition was shown to be

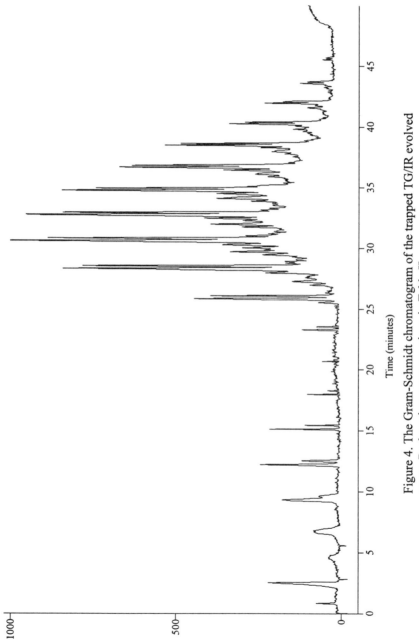

Figure 4. The Gram-Schmidt chromatogram of the trapped TG/IR evolved gases. Peak assignments are given in Table II.

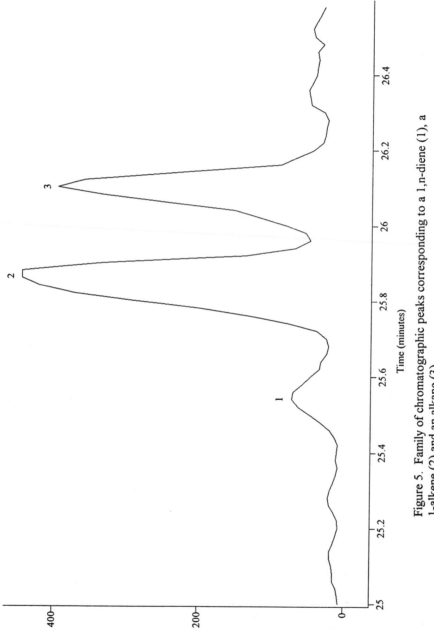

Figure 5. Family of chromatographic peaks corresponding to a 1,n-diene (1), a 1-alkene (2) and an alkane (3).

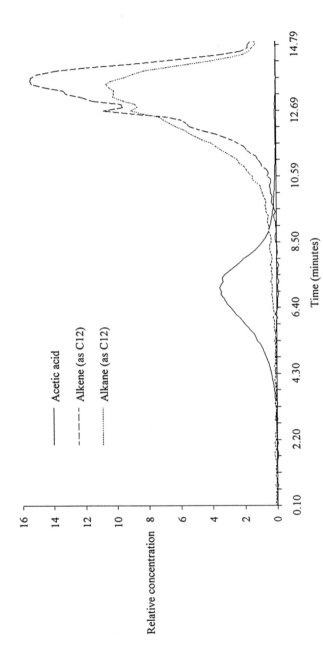

Figure 6. Curve-fit of the pure spectra of acetic acid, dodecane and dodecene obtained by GC/IR to the time slices obtained by TG/IR versus temperature.

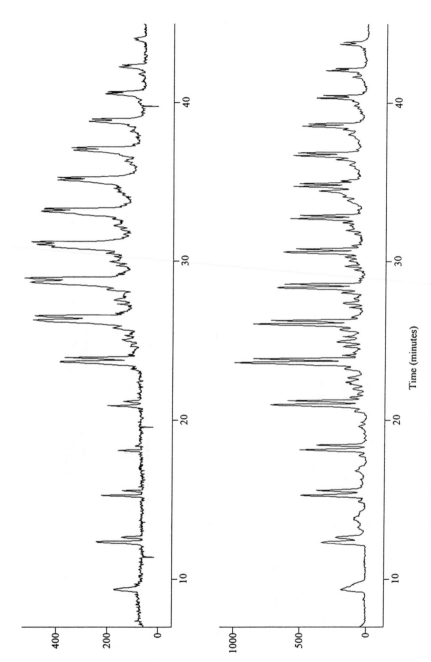

Figure 7. Gram-Schmidt chromatograms for nominally 12% and 40% vinyl acetate content in EVA copolymers.

much more complex than simply producing methane, ethylene and 1-butene, generating all volatile hydrocarbon chains from C_8 to C_{26}. These hydrocarbons were primarily in three moieties: a 1,n-terminal diene; a saturated hydrocarbon and a 1-alkene. The latter two functionalities dominated the evolved hydrocarbons.

In the curve fitting calculation, the saturated hydrocarbon appears to be produced at a lower temperature than its corresponding olefin. As to whether this was due to a shortening average chain length or to greater alkene production at higher temperature cannot be determined by this experiment. Further examination of the data using principal components regression is planned.

As the percentage of vinyl acetate in the samples increased, a shift in the hydrocarbon concentrations to shorter chain lengths was observed. This may indicate the cleaving of a C-C bond adjacent to the C=C bond, eliminating a saturated hydrocarbon and leaving a terminal alkene. Subsequent cleavage of the remaining terminal alkene adjacent to another C=C bond of the polyolefin may produce either a 1,n-diene (i.e., the bond breaks away from the terminal alkene) leaving a saturated terminal methyl group, or a 1-alkene again leaving a terminal alkene.

Acknowledgments

The author is grateful to Drs. Koichi Nishikida, David Schiering and Richard Spragg of The Perkin-Elmer Corporation for their helpful discussions and to Dr. Gregory McClure of Perkin-Elmer for confirming the chromatographic assignments by mass spectrometry and his suggestions in the preparation of this manuscript.

Trademarks

Tenax is a registered trademark of the Enka Research Institute. Perkin-Elmer is a registered trademark and AutoSystem is a trademark of The Perkin-Elmer Corporation.

Literature Cited

1. Wieboldt, R.C.; Adams, G.E.; Lowry, S.R., Rosenthal R.J., *Am. Lab.*, **1988**, *20(1)*, 70.
2. Van Dyke, J.D.; Kasperski, K.L. *J. Polym. Sci. Part A: Polym. Chem.*, **1993,** *31*, 1807.
3. Schild, H.G. *J. Polym. Sci. Part A: Polym. Chem.*, **1993**, *31*, 1629.
4. Maurin, M.B.; Dittert, L.W.; and Hussain, A.A. *Thermochimica Acta*, **1991**, *186*, 97.
5. Bowley, B.; Hutchinson, E.J.; Gu, P.; Zhang, M.; Pan, W.; Nguyen, C. *Thermochimica Acta*, **1992**, *200*, 309.
6. Solomons, T.W.G. *Organic Chemistry*, John Wiley and Sons: New York, NY, 1978, p. 185.

RECEIVED July 26, 1994

Chapter 9

Thermogravimetric Analysis—IR Spectroscopy: An Important Technique for the Study of Polymer Degradation

Charles A. Wilkie and Martin L. Mittleman[1]

Department of Chemistry, Marquette University, Milwaukee, WI 53233

The combined technique of thermogravimetric analysis - infrared spectroscopy, TGA/IR, provides very useful information to enable the understanding of the degradation scheme of a polymer. In addition to the usual weight loss information that is produced from the TGA portion of the experiment, the infrared spectra that are obtained permit temporal resolution of the gases that are evolved from the degrading polymer. In this paper, several systems that have been studied in these laboratories are reviewed and some of the mechanistic information that has been obtained is elucidated.

It is important to understand the thermal degradation pathways of polymers and the effect of additives on this degradation in order to either accelerate or retard the degradative process. In some cases, such as a landfill, it is advantageous to accelerate degradation while in other applications, such as in processing or reducing flammability, it would be desirable to retard it.

Thermal degradation of a polymer at high temperatures is usually complete and only low molecular weight fragments are obtained; this makes it difficult to infer a degradation pathway. Considerably more mechanistic information is available if degradation is carried out at modest, controlled temperatures and heating rates. Typical means of obtaining this information are either to perform thermolysis reactions in sealed tubes to collect and identify the products or to perform a real-time analysis in which the products are analyzed "on the fly." The latter approach can be accomplished by pyrolysis gas chromatography in which thermolysis is carried out at a controlled temperature and the products separated by GC and identified by mass spectrometry.

[1]Current address: Research Institute, University of Dayton, Dayton, OH 45469

0097–6156/94/0581–0116$08.00/0

Another effective technique is the use of thermogravimetric analysis coupled to an analytical method that will permit the identification of the products, such as mass spectrometry or infrared spectroscopy.   The theme of this paper is to demonstrate the utility of thermogravimetric analysis coupled to infrared spectroscopy, TGA/IR, to help understand the thermal degradation pathway of the polymer.   An important feature of this technique is that the evolved gases can be identified in sequential order and a specific gas may be associated with a specific weight loss.   In contrast to pyrolysis GC, in which all the gases produced by heating to a given temperature are separated and analyzed as a batch; TGA/IR offers the great advantage of sequentially identifying the gases.

We have been interested for some time in the effect of various additives upon the thermal degradation of poly(methyl methacrylate), PMMA. If one can understand the reactivity relationship between polymer and additive, it may be possible to design an additive to either accelerate or hinder degradation.   This paper will review the work that has been performed in our laboratories at Marquette University on this topic.

EXPERIMENTAL

Two different interfaces were used for TGA/IR.   The first employs a thermogravimetric analyzer supplied by PL Thermal Sciences coupled to a Bio-Rad FTS-60 Fourier Transform Infrared Spectrometer.   The second utilizes a Cahn thermogravimetric analyzer coupled to a Mattson Instrument Fourier Transform Infrared spectrometer.   For the first system the total gas flow passes continually through the infrared cell while in the second a sniffer tube extends into the TGA sample cup and admits only some of the gases into the analyzer. The latter result in less dilution of the degradation products by the TGA purge gas. For the first system, sample sizes ranging from 1 to 5 mg were used while, in the second, sample sizes were near 40 mg. In both cases the heating rate was 20°C per minute with a 30 - 50 cc/min inert gas purge of N_2 or Ar. Evolved gases were transferred to a heated 10 cm gas IR cell by a heated quartz transfer line.

Blends of PMMA with various additives and copolymers of methyl methacrylate with 2-sulfoethyl methacrylate (2-SEM) are discussed in this paper. Blends of transition metal salts and PMMA were prepared by dissolving the salt in a suitable solvent and combining this with a PMMA solution in chloroform. Homogeneous blends were obtained after evaporation of the solvents. The blend of Nafion-H and PMMA was obtained by pouring a chloroform solution of PMMA onto Nafion-H film and drying at room temperature.   The copolymers were obtained by standard techniques.

RESULTS AND DISCUSSION

The thermal degradation of PMMA has been studied by many workers. Recently two different research groups have focused new attention on this important polymer. Kashiwagi *et al.* (*1-4*) have implicated weak links in the polymer as the principal site of degradation. They have observed a three step degradation process and attributed these to the presence of head-to-head linkages in the polymer, end group unsaturation, and random scission. They suggested that side chain cleavage of the carbomethoxy group occurs after random scission of the polymer chain. Manring (*5-8*) proposes that weak links in the polymer backbone are less important and that degradation is initiated by cleavage of the pendant carbomethoxy group.

The thermal degradation of PMMA in the presence of additives has been studied by the McNeill group at the University of Glasgow and this group at Marquette University. McNeill and coworkers have examined the effect of silver acetate (*9*), ammonium polyphosphate (*10*), and zinc bromide (*11,12*) on the degradation of PMMA while the Marquette group has studied red phosphorus (*13,14*), $(PPh_3)_3RhCl$ (*15,16*), Nafion-H (*17*), $MnCl_2$ (*18*), $CrCl_3$ (*19*), $FeCl_2$, $FeCl_3$, $CuCl_2$, CuCl, and $NiCl_2$ (*20*), copolymers of MMA with 2-sulfoethyl methacrylate (*21*), and phenyltin chlorides (*22*). In this laboratory we have used both sealed tube reactions and TGA/IR techniques to probe the degradation pathways of the polymer. There is a significant difference between these approaches because in the sealed tube the products of degradation are retained within a constant volume and may undergo secondary reactions while in a TGA experiment they are flushed out with the purge gas and the possibility of recombination is greatly diminished.

Degradation of Nafion-H (*17*). Nafion-H is a DuPont product consisting of a tetrafluoroethylene backbone with pendant sulfonic acid groups. The TGA curve for Nafion-H shows that degradation proceeds in essentially a single step. About 5% weight is lost between 35 and 280°C with the product being water and sulfur dioxide. It is known that water is tenaciously retained by Nafion; vacuum-dried films still contain 2.7% water. Thus water lost up to 280°C results from desorption not degradation. Sulfur dioxide, on the other hand, is a thermal degradation product, clearly indicating that scission of the C-S bond is the first step in the degradation. Homolytic cleavage of the C-S bond produces a carbon-based radical and an SO_3H radical. This is followed by decomposition of the SO_3H radical to produce SO_2 and an OH radical. Subsequent reactions depend on the degradation of the carbon-based radical or the interaction of the hydroxy radical with polymer. At higher temperatures, 280 - 355°C, evolution of SO_2 continues and SiF_4, CO, HF, and carbonyl fluorides begin to appear. The carbonyl fluorides absorb at a unique position in the infrared, 1957 cm^{-1} and 1928 cm^{-1} and are easy to identify. In the highest temperature region, 355 - 560°C, the amount of SO_2 is greatly diminished and the major absorbances are due to SiF_4, HF, carbonyl fluorides, and C-F stretching vibrations. SiF_4 results from the reaction of product HF with the quartz transfer assembly. A scheme to account for this degradation is shown below as Scheme 1.

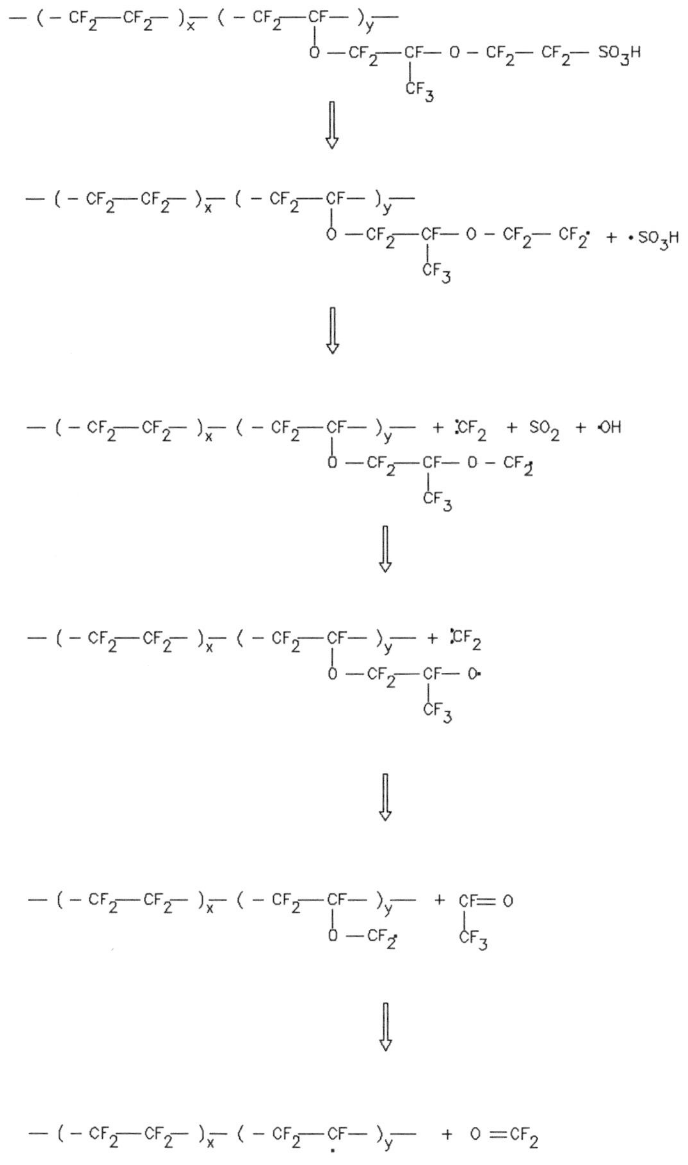

$$— (- CF_2—CF_2—)_{\overline{x}} \, (- CF_2—CF—)_{\overline{y}}—$$
$$O —CF_2—CF— O - CF_2— CF_2— SO_3H$$
$$CF_3$$

⇓

$$— (- CF_2—CF_2—)_{\overline{x}} \, (- CF_2—CF—)_{\overline{y}}—$$
$$O —CF_2—CF— O - CF_2— CF_2 \cdot \; + \cdot SO_3H$$
$$CF_3$$

⇓

$$— (- CF_2—CF_2—)_{\overline{x}} \, (- CF_2—CF—)_{\overline{y}}— \; + \; {}^{\cdot}CF_2 \; + \; SO_2 \; + \; {}^{\cdot}OH$$
$$O —CF_2—CF— O - CF_2$$
$$CF_3$$

⇓

$$— (- CF_2—CF_2—)_{\overline{x}} \, (- CF_2—CF—)_{\overline{y}}— \; + \; {}^{\cdot}CF_2$$
$$O —CF_2—CF— O\cdot$$
$$CF_3$$

⇓

$$— (- CF_2—CF_2—)_{\overline{x}} \, (- CF_2—CF—)_{\overline{y}}— \; + \; CF= O$$
$$O —CF_2 \qquad CF_3$$

⇓

$$— (- CF_2—CF_2—)_{\overline{x}} \, (- CF_2—CF—)_{\overline{y}}— \; + \; O =CF_2$$

Scheme 1

When Nafion-H is degraded in a sealed tube at 375°C, the observed products consist of SO_2 and SiF_4. The sealed tube experiment provides little information relating to mechanistic details of the degradation pathway. The carbonyl fluorides are probably present but, since they are in low concentration, they are not detected. Degradation of a blend of Nafion-H and PMMA (*17*). Degradation of a Nafion-H/PMMA blend does not occur in a single step, but shows at least three stages of degradation and yields a 10% residue, non-volatile at 600°C. In the first stage, 120 - 265°C, evolved gases are water and chloroform, both resulting from desorption. Actual degradation does not begin until 265°C. In the degradation of Nafion-H alone SO_2 is observed at lower temperatures. Apparently PMMA retards cleavage of the C-S bond. A small amount of degradation occurs between 265 and 360°C, evolving sulfur dioxide, CO_2, CO, and methyl methacrylate. Major degradation occurs between 360 and 450°C. Monomeric methyl methacrylate is the primary product in this region; it is first observed at 270°C and grows in intensity until 415°C and then significantly decreases. Other products include SO_2, CO_2, CO, and SiF_4. Notably absent are the carbonyl fluorides. Between 430 and 575°C the products are similar to those seen in the degradation of Nafion-H alone with two important differences. The fluorinated species observed in Nafion-H at 350°C are not seen in the blend until 450°C. Carbonyl fluorides are major products at 280°C in Nafion-H alone. In the blend these are observed as only very minor products above 480°C. Scheme 2 accounts for these observations; the interaction between the Nafion-H radical and PMMA probably occurs by a radical combination process.

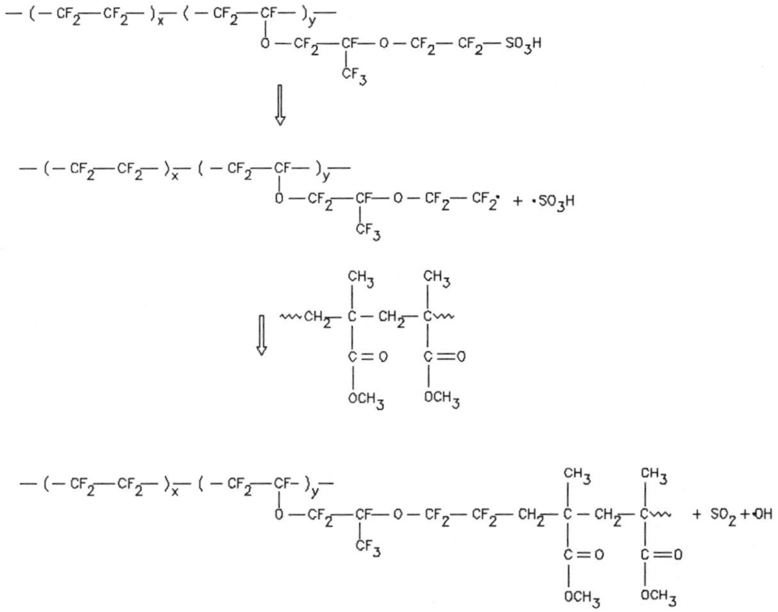

Scheme 2

Degradation of poly(2-sulfoethyl methacrylate) and methyl methacrylate copolymers (*21*). The TGA curve for poly(2-SEM) is shown in Figure 1. Degradation occurs in essentially a single step beginning near 200°C and ending near 330°C with the formation of 16% non-volatile residue at 600°C. Gases initially evolved are water, carbon dioxide, and carbon monoxide. At 250°C, evolution of sulfur dioxide and ethylene are noted. Water loss continues until 300°C, indicating that it originates from decomposition rather than desorption. Conspicuous by their absence are carboxylic esters. This indicates that monomer is not produced during the degradation. This is very surprising for a methacrylate; only poly(methacrylic acid), of all the methacrylate polymers, produces a residue and this is only 2% (*23*).

Several copolymers of 2-SEM with methyl methacrylate were also prepared and analyzed by TGA/IR. TGA curves for this family of copolymers are shown in Figure 2. Little or no monomeric methyl methacrylate is obtained from copolymers which contain 40 or more wt% 2-SEM. The onset of degradation of the 2-SEM portion of the copolymer increases with increasing MMA content. A plot of non-volatile residue versus 2-SEM content of the copolymers indicates more char than would be expected from the 2-SEM content alone. This suggests that PMMA is also involved in char formation, this is illustrated in Figure 3. The mechanistic schemes for the degradation of poly(2-sulfoethyl methacrylate) and its copolymers are shown as Schemes 3 and 4 respectively. In Scheme 4 diradicals are indicated. These probably do not arise in a single step; the important idea is that both side groups are eliminated.

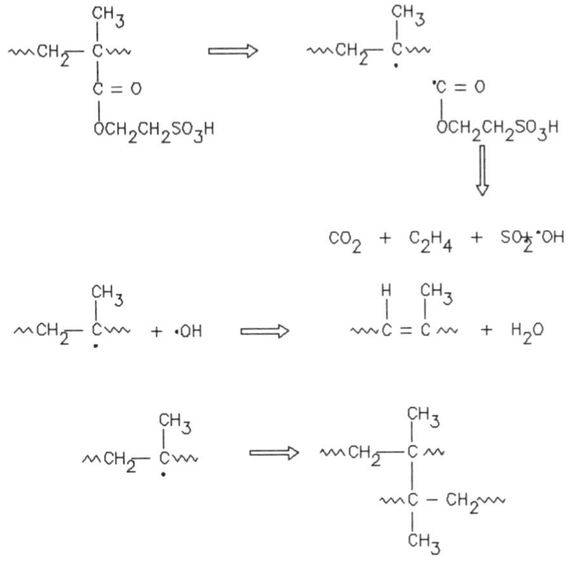

Scheme 3

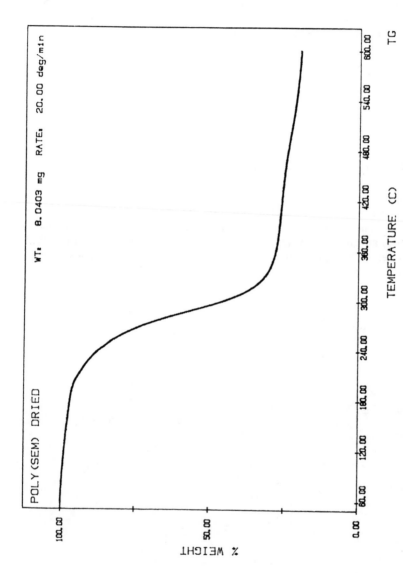

Figure 1. TGA curve for poly(2-SEM). Rate is 20°C per minute. (Reproduced with permission from ref. 21. Copyright 1993 Elsevier Applied Sciences.)

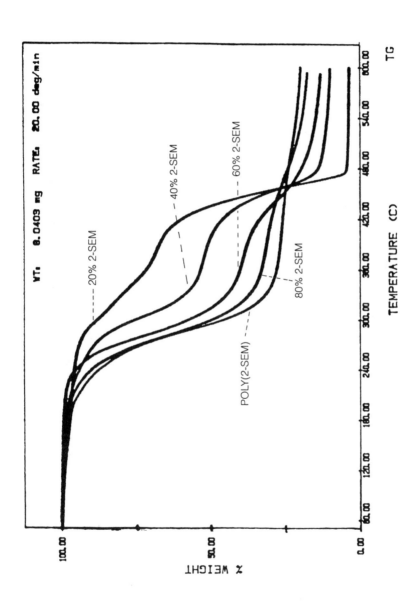

Figure 2. TGA curves for the copolymers of MMA with 2-SEM. Rate is 20°C per minute. (Reproduced with permission from ref. 21. Copyright 1993 Elsevier Applied Sciences.)

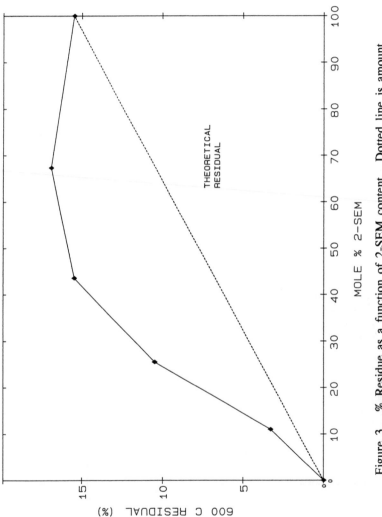

Figure 3. % Residue as a function of 2-SEM content. Dotted line is amount expected based upon 2-SEM content, solid line is actual amount found. (Reproduced with permission from ref. 21. Copyright 1993 Elsevier Applied Sciences.)

Comparison of the physical blend, Nafion-H/PMMA, with the chemically combined copolymers. The degradation of Nafion-H/PMMA blends and 2-SEM/MMA copolymers both show significant char formation, indicating that the methyl methacrylate units of both interact with the sulfonic acid groups. As illustrated in the Schemes, the process by which this occurs is different but the end result is that MMA is incorporated into the char rather than simply producing monomer as is seen in PMMA degradation.

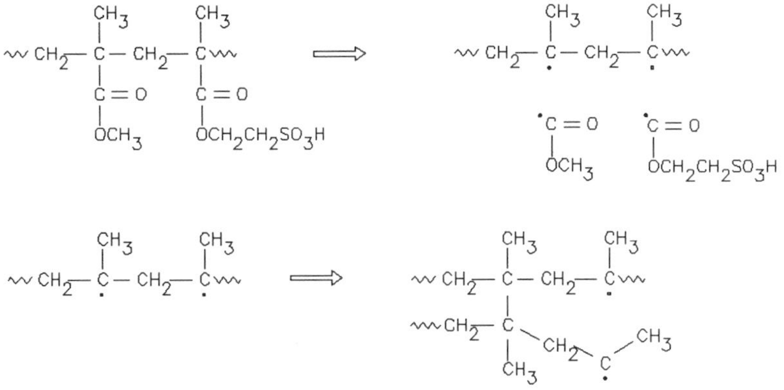

Scheme 4

Degradation of PMMA in the presence of transition metal compounds. McNeill (*11,12*) has established that the initial step when a transition metal halide is combined with PMMA is coordination of the metal ion to the carbonyl oxygens of the polymer. This is usually followed by the loss of methyl halide and the formation of a metal polymethacrylate. TGA/IR has provided interesting means to probe these systems to compare results using manganese chloride (*18*) and chromium (III) chloride (*19*) as PMMA additives. Weight loss data and identification of the evolved gases for both additives are reported below in Table I.

It is evident that the interaction of the additive with PMMA in the two cases is quite different. Water is again desorbed at low temperatures as a result of the hygroscopic nature of the transition metal salts. Methyl chloride is evolved from PMMA combined with manganese chloride but not from that combined with chromium chloride. The initial step in both cases is believed to be coordination of the metal ion to the carbonyl oxygens. For the manganese salt, the next step is loss of methyl chloride but not in the case of the chromium salt. At elevated temperatures $CrCl_3$ loses a chlorine atom with formation of chromium (II) chloride. The chlorine atom abstracts a hydrogen atom with transfer of a methyl from the ester group to the main chain.

Table I

TGA/IR Data for $MnCl_2$/PMMA and $CrCl_3$/PMMA

$MnCl_2$/PMMA

Temperature	Identity of products	Weight loss
100-145°C	H_2O	7.5%
145-215°C	monomer	9.5%
215-340°C	monomer, CH_3Cl	10.5%
340-455°C	CO_2, CO, CH_4, HCl	9%
455-630°C	HCl	5%
residue		58.5%

$CrCl_3$/PMMA

Temperature	Identity of Products	Weight Loss
100-250°C	H_2O, HCl, monomer	22%
250-500°C	monomer, HCl, CO_2, CO, CH_4	62%
residue		16%

The TGA curves for blends of PMMA with $FeCl_2$ and $FeCl_3$ are similar and the same gases are evolved. For the PMMA/$FeCl_2$ blend, the degradation occurs in three steps. Between 135 and 200°C, 13% of the sample volatilizes. The evolved gases consist of methanol, MMA, CO_2, CO, and methyl chloride. An additional 15% of the sample mass is lost between 220 and 270°C and the same gases are observed. 24% of the sample is evolved between 270 and 600°C. HCl begins to appear at 450°C and methyl chloride is absent at this temperature. Above 500°C only HCl and methane may be observed. Figure 4 shows a stacked plot of infrared spectra in this region and illustrates the appearance of these gases. The blend of $FeCl_3$ with PMMA gives very similar results. Monomer, CO, and CO_2 evolution begins near 200°C; near 300°C methyl chloride appears. Monomer is greatly diminished at 450°C while CO_2 and CO evolution continue. HCl appears initially near 480°C but methyl chloride is no longer seen at this temperature. At temperatures greater 480°C, HCl is essentially the only product of the degradation. The appearance of methyl chloride indicates that both blends initially form a complex between the metal ion and the carbonyl oxygens, followed by the loss of methyl chloride, and the formation of an iron salt of the polymer.

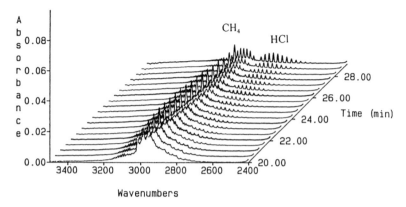

Wavenumbers

Figure 4.   Stacked plot of infrared spectra in the region between 3400 and 2400cm^{-1} for the degradation of the PMMA/FeCl$_2$ blend. 20 minutes corresponds to 450°C and the temperatures increases at a rate of 20°C per minute.

For the blend with CuCl$_2$, CO$_2$ and monomer begin to appear at 260°C with CO$_2$ being dominant. As the temperature increases, CO$_2$ and monomer release continue and CO begins to appear.   At 420°C HCl is obvious, monomer diminishes, and CO$_2$ and CO release continue. Near 480°C only HCl is evolved. For the CuCl blend both monomer and CO$_2$ appear near 200°C.   As the temperature is increased these grow in intensity and CO begins to appear near 400°C. Monomer grows in intensity, then diminishes and is essentially gone by 530°C. The gases at 530°C consist of HCl, CO$_2$, CO, and CH$_4$. Methyl chloride is never observed in the degradation of this blend; this is likely a result of the lack of coordination of the Cu$^+$ ion to the carbonyl oxygens of the PMMA.

The degradation of the blend of nickel chloride with PMMA begins near 150°C with the evolution of monomer and smaller amounts of CO$_2$. CO is first observed at 360°C and HCl begins to be evolved at 400°C. By the time the temperature has reached 440°C, HCl is dominant and all other gases are diminishing.

Two steps are important to understand the effect of transition metal chlorides on the degradation of PMMA: coordination and loss of chlorine atoms. In the case of copper (I) ion, no coordination occurs and there is no effect on the degradation. For iron (II), iron (III), and manganese (II) the metal-chlorine bond is relatively weak and easily cleaved and methyl chloride is produced. The M-Cl bonds in nickel (II) chloride and copper (II) chloride are stronger so cleavage is more difficult and methyl chloride is not obtained (*24*).

CONCLUSION

TGA/IR studies of polymer degradation provide very useful information about the degradation process and permit one to learn details about the degradation pathway that cannot be obtained by sealed tube reactions or pyrolysis gas chromatography. It seems that the presence of methyl halide as a degradation product when transition metal halides are combined with PMMA is very indicative of the type of degradation pathway. TGA/IR is not the solution to all problems in the study of polymer degradation and other techniques are required in order to completely identify the complete reaction scheme; however, this technique is a very valuable first experiment in order to delineate the reaction mechanism and identify the major products.

LITERATURE CITED

1.   Kashiwagi, T.; Harita, T.; and Brown, J.E.; *Macromol.*, **1985**, *18*, 131.
2.   Harita, T.; Kashiwagi, T.; and Brown, J. E.; *Macromol.*, **1985**, *18*, 1410.
3.   Inabi, A.; Kashiwagi, T.; and Brown, J. E.; *Polym. Degrad. Stab.*, **1988**, *21*, 1.
4.   Kashiwagi, T.; Inaba, A.; and Hamins, A.; *Polym. Degrad. Stab.*, **1989**, *26*, 161.
5.   Manring, L.; *Macromol.*, **1988**, *21*, 528.
6.   Manring, L.; *Macromol.*, **1989**, *22*, 2673.
7.   Manring, L.; *Macromol.*, **1989**, *22*, 4652.
8.   Manring, L.; *Macromol.*, **1989**, *24*, 3304.
9.   Jamieson, A. and McNeill, I. C.; *J. Polym. Sci.: Polym. Chem. Ed.*, **1978**, *16*, 2225.
10. Camino, G.; Grassie, N.; and McNeill, I. C.; *J. Polym. Sci.: Polym. Chem. Ed.*, **1978**, *16*, 95.
11. McNeill, I. C.; and McGuiness, R. C.; *Polym. Degrad. Stab.*, **1984**, *9*, 167.
12. McNeill I. C. and McGuiness, R. C.; *Polym. Degrad. Stab.*, **1984**, *9*, 209.
13. Wilkie, C. A.; Pettegrew, J. W.; and Brown, C. E.; *J. Polym. Sci., Polym. Lett. Ed.*, **1981**, *19*, 409.
14. Brown, C. E.; Wilkie, C. A.; Smukalla, J.; Cody, Jr., R. B.; and Kinsinger, J. A.; *J. Polym. Sci., Polym. Chem. Ed.*, **1986**, *24*, 1297.
15. Sirdesai, S. J. and Wilkie, C. A.; *J. Appl. Polym. Sci.*, **1989**, *37*, 863.
16. Sirdesai, S. J. and Wilkie, C. A.; *J. Appl. Polym. Sci.*, **1989**, *37*, 1595.
17. Wilkie, C. A.; Thomsen, J. R.; and Mittleman, M. L.; *J. Appl. Polym. Sci.*, **1991**, *42*, 901.
18. Wilkie, C. A.; Leone, J. T.; and Mittleman, M. L.; *J. Appl. Polym. Sci.*, **1991**, *42*, 1133.
19. Beer, R. S.; Wilkie, C. A.; and Mittleman, M.L.; *J. Appl. Polym. Sci.*, **1992**, *46*, 1095.
20. Chandrasiri, J. A.; Roberts, D; and Wilkie, C. A.; *Polym. Degrad. Stab.*, in press.
21. Hurley, S. L.; Mittleman, M. L.; and Wilkie, C. A.; *Polym. Degrad. Stab.*, **1993**, *39*, 345.
22. Chandrasiri, J. A.; and Wilkie, C. A.; *Polym. Degrad. Stab.*, in press.
23. McNeill, I. C.; *Develop. Polym. Degrad.*, **1987**, *7*, 1.
24. Berg, R. A.; and Sinanoglu, O.; *J. Chem. Phys.*, **1960**, *32*, 1082.

RECEIVED July 26, 1994

Chapter 10

Applications of Thermogravimetry–Fourier Transform IR Spectroscopy in the Characterization of Weathered Sealants

Ralph M. Paroli[1] and Ana H. Delgado

Institute for Research in Construction, National Research Council of Canada, Ottawa ON K1A OR6, Canada

The combination of thermogravimetry-Fourier transform infrared spectroscopy (TG-FTIR) was used to study the effects of accelerated weathering on silicone and polyurethane sealants. All evolved gases from the TG are sent to an FTIR spectrometer using a heated transfer line. The results demonstrate that this technique can be useful in identifying the decomposition products of construction sealants. It is relatively simple and can be adapted to most TG and FTIR combinations. Care must be taken that no leaks in the transfer line occur since it could lead to peaks appearing in the infrared spectrum but not appearing in the TG curve (e.g., oxidation of polymer backbone). This technique can be used to monitor the changes in chemical composition due to aging or weathering.

Sealants are used to seal cracks and joints in window frames or between panels, to prevent rain, air and dust from passing through the joint and even to improve the thermal performance of a wall. Sealants are available in a non-cured, pourable or extrudable state for easy application. Upon curing, they are transformed into a solid elastomeric material. Sealants must be deformable, have good recovery properties and should have good overall elastic properties. They should be durable and should not be affected by the environment to which they will be subjected. The general composition of a sealant consists of a base polymer and additives. Sealants can be classified based on their chemical composition. Two of the most widely used groups in the construction industry, are silicones and polyurethanes (1).

---

[1]Corresponding author

0097–6156/94/0581–0129$08.00/0
Published 1994 American Chemical Society

$$\left[\begin{array}{c} Me \\ | \\ HO - Si - O - \\ | \\ Me \end{array} \left[\begin{array}{c} Me \\ | \\ Si - O \\ | \\ Me \end{array} \right]_x \begin{array}{cc} Me & O \\ | & || \\ - Si - (OCCH_3)_2 \\ \end{array} \right]$$

Base Polymer for Silicone Sealant

$$\begin{array}{cc} H & O \\ | & || \\ - N - C - O - \end{array}$$

Urethane Linkage

During natural or accelerated weathering the chemical properties of a material change. These changes can be studied by different analytical techniques such as thermal analysis and Fourier-transform infrared spectroscopy. Thermal analysis includes a wide range of techniques such as thermogravimetry (TG), differential thermal analysis (DTA), and simultaneous thermal analysis (STA or TG/DTA or TG/DSC) (2). TG provides information on the thermal stability and composition of the polymeric material by measuring change in mass as a function of temperature and/or time (3). FTIR spectroscopy encompasses techniques such as photoacoustic (PAS-FTIR), diffuse reflectance (DRIFT-FTIR), attenuated total reflectance (ATR-FTIR) and microscopy-FTIR.

TG is gaining acceptance as a method for the compositional analysis of vulcanizates and especially in the characterization of construction materials (4-7). It is also widely used for studying degradation mechanisms, predicting service lifetime and measuring thermal decomposition. More recently it has been used for the characterization of construction sealants (8,9). TG yields quantitative data such as weight loss. However, to identify the gaseous components evolving during a TG experiment it is necessary to couple the TG to another analytical device such as an infrared spectrometer (TG-FTIR) (10-18). This is crucial in the field of polymeric building materials because they contain a wide variety of additives such as plasticizers, antioxidants, antiozonants, etc. The added amounts sometimes make it difficult to analyze by FTIR alone and it is of paramount importance to identify which components, if any, are changing with aging. The combination of TG-FTIR is ideal since all evolved gases from the TG are sent to an FTIR spectrometer using a heated transfer line. The FTIR spectrum is acquired and compared to spectra in data banks. Through this hyphenated technique a more complete characterization of thermal stability and decomposition products can be achieved. Moreover, this approach not only shortens the analysis time but also permits the analysis to be performed on the same specimen, thus minimizing experimental errors (due to sample inhomogeneity).

Experimental

Weathering (9). Free films of commercially available silicone and polyurethane construction sealants were placed back-to-back in a frame where the films were separated with an aluminum divider. All of the free films were loaded into a xenon-arc Weather-o-meter (Atlas Electric Devices Company) for weathering. One series of films was thus exposed to light and water while the other was exposed to water only. The total time of exposure was 8000 hours.

The weathering conditions used for the samples were as follows:

Irradience: 0.37 W m^{-2} nm
Black panel temperature: 63 °C
Relative humidity arc on: 50%
Light cycle: 3.5 hours on; 0.5 hours off
Specimen spray cycle (deionized water): 118 min. off; 20 min. on
Rack spray cycle: when arc off

Thermogravimetric and Fourier transform infrared analysis (TG-FTIR). TG-FTIR studies of unweathered and weathered sealants were performed using a Seiko Simultaneous Thermal Analyzer (STA) model TG/DTA320 connected to a Nicolet 800 Fourier-transform infrared spectrometer. Since this appears to be the first time that this combination of equipment has been used, a detailed description is given. The data from the TG/DTA320 were recorded and processed with a SSC5200H disk station. The STA was equipped with a vacuum furnace tube available from Seiko Instruments Inc., USA. The gases emanating from the sample were carried from the furnace using a purge gas. This furnace was coupled to an Nicolet 800 Fourier-transform infrared spectrometer using a TG interface kit supplied by Nicolet Instrument Inc. The kit consisted of a Teflon adapter connected, a heated transfer line, and a glass cell encased in an insulated chamber. Both the transfer line and chamber are heated to a determined temperature. It is important to note that both the STA and the FTIR are independently controlled and therefore, the controller programmes must be initiated separately. A schematic of the setup is shown in Figure 1.

In this paper, two silicone and two polyurethane unweathered and weathered sealants (6000 hours exposure time) were analyzed by TG-FTIR. Approximately 15 mg of sample was placed in the TG pan and heated from 40 °C to 900 °C at 10 °C/min under a nitrogen atmosphere (100 mL/min). The evolved gas was transferred from the STA to the FTIR by a heated transfer line maintained at 240 °C. The FTIR glass cell was also set at 240 °C. The FTIR detector was DTGS. Data collection on the FTIR (using the Nicolet SID software) was started when the sample temperature on the STA reached 100 °C. Spectra were collected at 8 cm^{-1} and a mirror velocity of 30. Samples were ratioed against a background collected under the same experimental conditions as the sample but prior to any gas evolution. Spectra are baseline-corrected.

Results and Discussion

A typical plot obtained (from the FTIR spectrometer) by coupling the TG to an FTIR is shown in Figure 2. This is commonly referred to as a Gram-Schmidt reconstruction (GSR) and is an indirect quantitative indicator of the infrared absorption of the evolving gases. A corresponding TG/DTG/DTA plot is given in Figure 3. As can be seen, the derivative weight loss (DTG) and the GSR both show two peaks. The difference in peak intensities has been attributed to the different absorptivities of the components (*10*). Although STA (TG/DTA and DTG) data for all sealant materials studied are available, only the DTG curves will

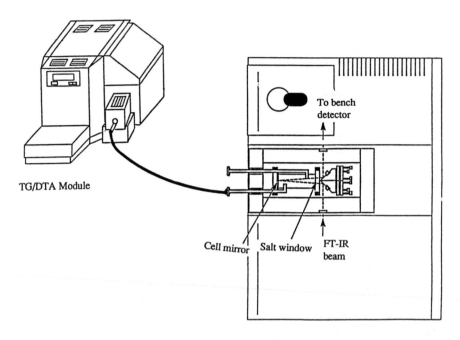

Figure 1. Schematic of STA (TG/DTA)-FTIR coupling.

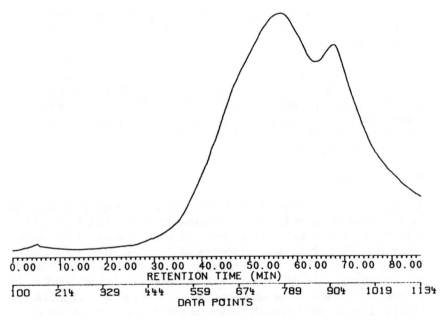

Figure 2. Plot of a typical Gram-Schmidt reconstruction (GSR).

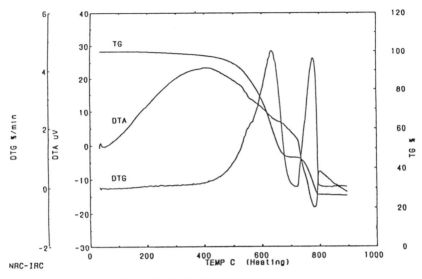

Figure 3. Typical TG/DTG/DTA plot.

be used because, in this case, the DTA data did not provide further information regarding the degradation behaviour of the sealants.

Silicone Sealants. The DTG traces for the control and weathered samples from the silicone series S1 and S2 are displayed in Figures 4a and 4b. As can be seen, both series lose weight in two steps. The total weight loss was approximately 75%. The first weight loss occurs in the range of 200-700 °C and the other in the range of 700-900 °C. Although the control and exposed samples showed similar weight losses their DTG curves are slightly different. The two peaks of DTG curve for the control sample S1 show similar peak-height. However, in the samples exposed to UV-radiation and water, the peak due to the first weight loss in the sample exposed to UV-radiation and water becomes broader and a slight decrease in intensity is observed. The second peak becomes sharper with an increase in peak-height. The peak-height of the sample exposed to water not only shows a smaller peak for the first weight loss but also a shoulder between 450-550 °C. Contrary to the sample exposed to UV-radiation and water, the peak-height of the second peak decreases considerably. The DTG peaks for the S2 sample showed similar features as the S1 series, therefore they will not be discussed further.

The TG-FTIR spectra for S1 are displayed in Fig. 5(a-c). A comparison of the DTG curves with the FTIR spectra of the samples indicates that the first weight loss observed in the DTG curves corresponds to the decomposition of the polymer. A library search indicated that the main components given off can be attributed to poly(dimethylsiloxane) with an aromatic component. The absorption bands at 3017, 2968 and 2908 cm^{-1} are due to aromatic and aliphatic CH bonds. Absorption bands at 1266 cm^{-1} and 780 cm^{-1} correspond to Si-CH$_3$. The bands at 1084 and 1024 cm^{-1} are the due to Si-O-Si bonds. The relative intensities of the Si-O-Si peaks vary as the temperature increases perhaps due to different polymer backbone chain lengths. This could imply that different siloxane polymers are evolving as a

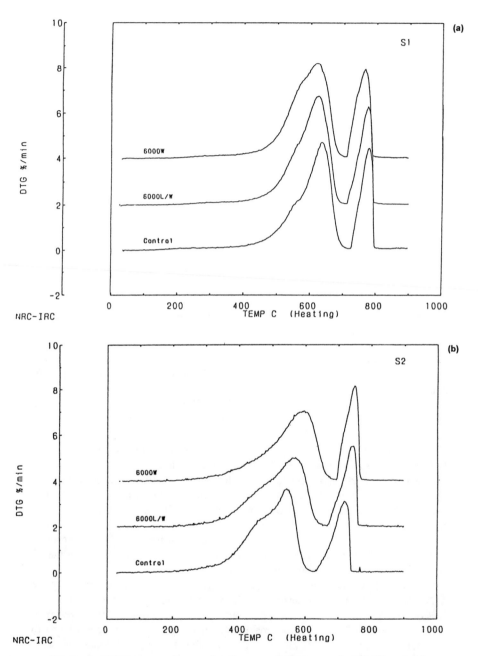

Figure 4. DTG curves of control and exposed silicone series (a) S1 and (b) S2.

S1 Control

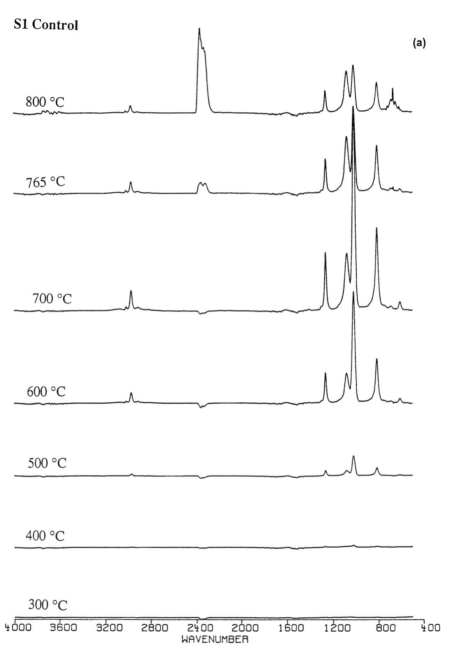

Figure 5. TG-FTIR spectra of silicone series S1; (a) control, (b) 6000 hours light and water exposure, and (c) 6000 hours water exposure only.

Continued on next page

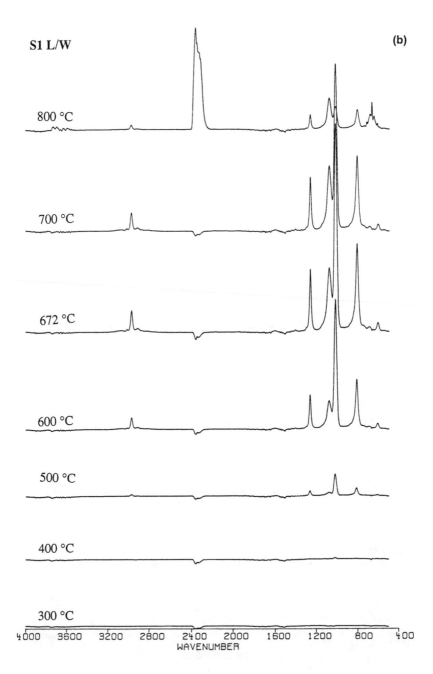

Figure 5. *Continued.* TG-FTIR spectra of silicone series S1; (a) control, (b) 6000 hours light and water exposure, and (c) 6000 hours water exposure only.

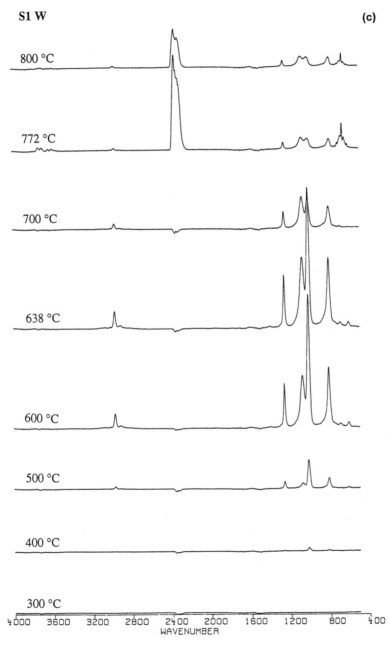

Figure 5. *Continued.* TG-FTIR spectra of silicone series S1; (a) control, (b) 6000 hours light and water exposure, and (c) 6000 hours water exposure only.

S2 800 °C

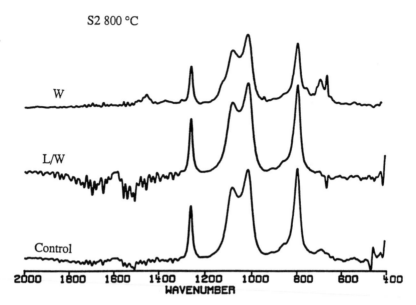

Figure 6. TG-FTIR spectra of silicone series S2 at 800°C.

function of temperature. Moreover, this relative intensity also varies for the different formulations, i.e., the Si-O-Si peak relative intensities were different for S1 and S2 (see Figure 6). The weight loss above 700 °C was found to be due to calcium carbonate as indicated by the large increase in the CO_2 absorption bands above 700 °C. The presence of calcium carbonate was also confirmed by the PAS-FTIR spectrum of a quenched sample.

Polyurethane Sealants. Similar observations were made for both polyurethane sealant samples. The main differences observed were due to differences in the formulations. It is for this reason that the discussion for this section will concentrate on only one sealant.

The DTG curves for the PU1 and PU2 control and exposed samples show two weight losses between 200 °C and 400 °C (see Figures 7a, 7b). This weight loss region is due to decomposition of the polymer. After weathering, this region changes dramatically for PU2 as the two resolved peaks merge and become unresolved. A small weight loss is observed at approximately 450 °C in the exposed and unexposed PU2 samples, while a major weight loss is observed at ~750 °C for PU1. This latter peak is most likely due to calcium carbonate.

The TG-FTIR spectra of evolved gases emanating from the PU2 samples are shown in Figures 8(a-c). It is interesting to note that HCl appears to be evolving during the decomposition of the material above 250 °C. This is easily identifiable by the rotational peaks occurring in the region between 3100 cm^{-1} and 2550 cm^{-1}. No HCl was detected in the other polyurethane sealant sample (see Figure 9). At

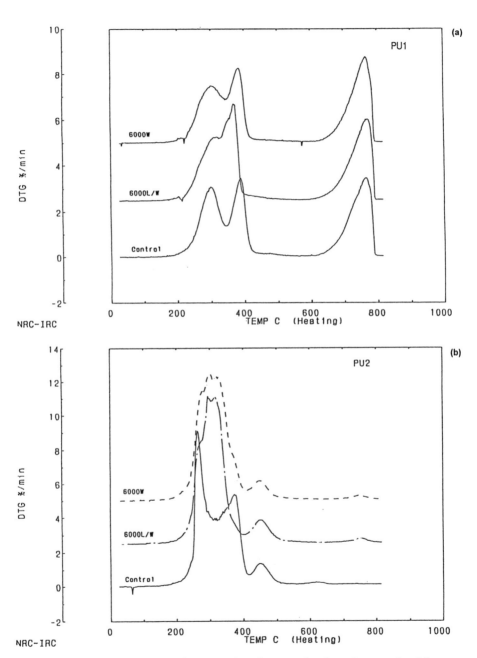

Figure 7. DTG curves for control and exposed polyurethane series (a) PU1 and (b) PU2.

PU2 Control

(a)

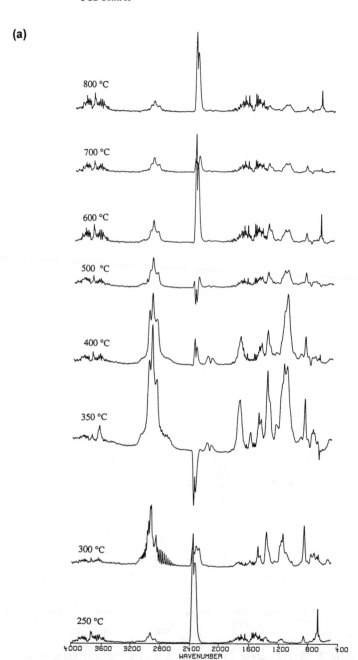

Figure 8. TG-FTIR spectra of polyurethane series PU2; (a) control, (b) 6000 hours light and water exposure, and (c) 6000 hours water exposure only.   *Continued on next page*

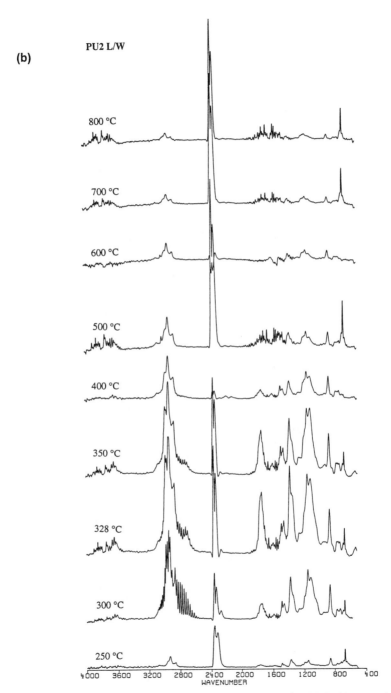

Figure 8. *Continued.* TG-FTIR spectra of polyurethane series PU2; (a) control,
(b) 6000 hours light and water exposure, and (c) 6000 hours water exposure only.

Continued on next page

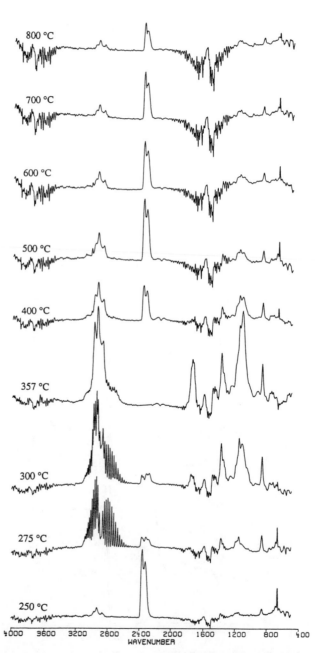

Figure 8. *Continued.* TG-FTIR spectra of polyurethane series PU2; (a) control, (b) 6000 hours light and water exposure, and (c) 6000 hours water exposure only.

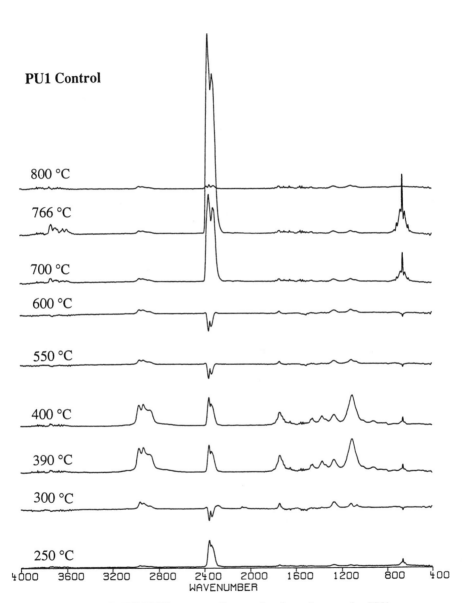

Figure 9. TG-FTIR spectra of control polyurethane series PU1.

higher temperatures, polymer peaks begin to be more prominent. These peaks can be attributed to urethane bonds. For example, the peak at ~1100 cm^{-1} is most likely due to polyether while the one at ~1700 cm^{-1} is a urethane carbonyl peak. After weathering, the same peaks are observed but the decomposition is initiated at lower temperatures.

The CO_2 region between 2400 cm^{-1} and 2340 cm^{-1} is also worth discussing. Figure 8 contains the spectra for the control and exposed samples of PU2 acquired between 250 °C and 800 °C. As can be seen, CO_2 peaks are positive between 250 °C and 350 °C but become negative at 350 °C for the control sample. The positive peaks at low temperatures are not due to the decomposition of calcium carbonate (a filler present in this sealant) since the temperature is too low. It is quite possible that at this temperature the polyurethane polymer is decomposing. A pyrolysis-FTIR study (19) has shown that for polyurethanes based on MDI, 1,4-butane diol and polyadipate three CO_2 events are associated with the thermal degradation of the polyurethane. The first is associated with the decomposition of the urethane links, the others are associated with the formation of cyclopentanone. The negative CO_2 peaks in Figure 8a are due to the presence of CO_2 in the background spectrum.

Another region of interest is between 2340 cm^{-1} and 2200 cm^{-1}. This region is associated with isocyanates (NCO). The two peaks around 2305 cm^{-1} and 2275 cm^{-1} (Figure 10) could be due to unreacted isocyanate groups (20-21) which upon aging or weathering will react further. It is interesting to note that these peaks are no longer present after 350 °C and that only very weak peaks were observed in the FTIR spectra of the weathered samples (Figure 11). It is also possible that one of these peaks originated from the decomposition of bis(2,6-diisopropyl)carbodiimide, an anti-hydrolysis additive, which upon decomposition yields isocyanates (22).

The region between 2180 cm^{-1} and 2110 cm^{-1} also showed some changes during the TG-FTIR experiment (Figure 11). This domain is usually associated with

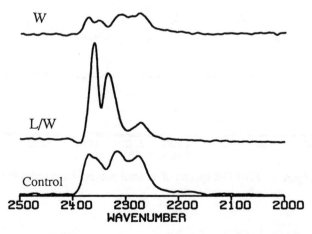

Figure 10. TG-FTIR spectra of control and exposed polyurethane series PU2 at 300 °C.

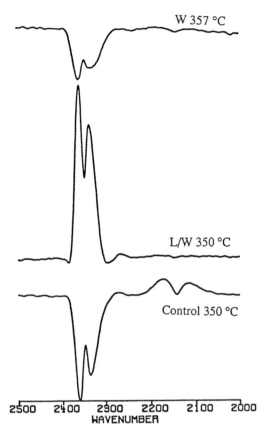

Figure 11. TG-FTIR spectra of control and exposed polyurethane series PU2 at 350 °C.

carbon monoxide which can be observed when the polymer backbone undergoes oxidation. It could also be due to carbodiimide (-N=C=N-) group which is present in polyurethanes. This region produced peaks which were extremely weak and difficult to detect. In this case, however, the peaks in this region are most likely due to CO since they only appeared when a small leak between the TG and the FTIR was detected (see Figures 12 and 13).

Conclusions

The results obtained from TG-FTIR demonstrate that this technique can be useful in identifying the decomposition products of construction sealants. It is relatively simple and can be adapted to most FTIR-TG combinations. Care must be taken that no leaks in the transfer line occur since it could lead to peaks appearing in the

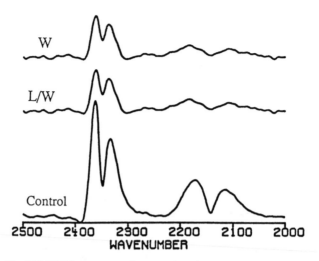

Figure 12. TG-FTIR spectra of control and exposed polyurethane series
PU2 at 400 °C.

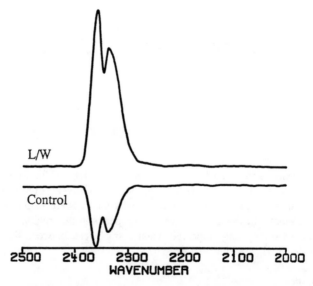

Figure 13. TG-FTIR spectra of control and exposed polyurethane series
PU2 at 400 °C (without leak in transfer line connection).

infrared spectrum which do not correspond to the TG curve (e.g., oxidation of polymer backbone). This technique can be used to monitor the changes in chemical composition due to aging. This information could then be used to modify the formulation and increase performance as well as the service-life of the material.

Literature Cited.

1. Feldman, D. Polymeric Building Materials; Elsevier Applied Science: New York, NY, 1989; pp. 391-408.
2. Farlling, M.S. *Rubber World.* **1988**, *197*, pp. 20-48.
3. Flynn, J.H. In *Thermal Analysis*: Mark H.F.; Bikales, N.M.; Overberger, C.G.; Menges, G.; *Encyclopedia of Polymer Science and Engineering*, 2nd Edition, John Wiley & Sons: New York, NY, Supplement Volume; 1989; pp. 690-723.
4. Sircar, A.K. *Rubber Chemistry and Technology*, **1991**, *65*. pp. 503-526.
5. Paroli, R.M.; Dutt, O.; Delgado, A.H.; Mech, M.N. *Thermochim. Acta.* **1991**, *182*, pp. 303-317.
6. Paroli, R.M.; Dutt, O.; Delgado, A.H; Stenman H. K. *J. of Materials in Civil Engineering*, **1993**, *5*, pp. 83-95.
7. Penn, J.J.; Paroli, R.M. *Proceedings of the Twenty-First North American Thermal Analysis Society Conference*, **1992**, *21*, pp. 612-617.
8. Paroli, R.M.; Delgado, A.H.; Cole, K.C. *Can. J. Appl. Spectros.*, **1994**, *39*, pp. 7-14.
9. Paroli, R.M; Delgado, A.H. *Proceedings of the American Chemical Society, Division of Polymeric Materials: Science and Engineering*, **1993**, *68,* pp. 334-335.
10. Wieboldt, R. C.; Lowry, S.R.; Rosenthal, R.J. *Mikrochim. Acta* [Wien], **1988**, *1*, 179-182.
11. Ramirez, E.; Sanchez, V.S.; Huerta M., B.H. *Revista de Plasticos Modernos*, **1992**, *433*, pp. 59-64.
12. Compton, D.A.C.; Johnson, D.J.; Mittlemen, M.L. *Research and Development*, **1989**, *April*, pp. 68-73.
13. Johnson, D.J.; McCarthy, W.J. *Proceedings of the Twenty-First North American Thermal Analysis Society Conference*, **1992**, *21*, p. 165
14. Schild, H.G. *J. Polym. Sci. Part A: Polym. Chem.*, **1993**, pp. 1629-1632.
15. Wilkie, C.A.; Mittleman, M.L. *Proceedings of the American Chemical Society, Division of Polymeric Materials: Science and Engineering*, **1993**, *69*, p. 134.
16. White, S.E.; M.L. McGrattan, M.L. *Proceedings of the American Chemical Society, Division of Polymeric Materials: Science and Engineering*, **1993**, *69*, pp. 135-136.
17. Johnston, D.J.; Stout, P.J.; Hill, S.L.; Krishnan, K. *Proceedings of the American Chemical Society, Division of Polymeric Materials: Science and Engineering*, **1993**, *69*, pp. 137-138.
18. Redfern, J.P.; Newbatt, P.H.; Larcey, P. *Proceedings of the American Chemical Society, Division of Polymeric Materials: Science and Engineering, Fall Meeting*, **1993**, *69*, pp. 144-145.
19. Davidson, R.G. *Mikrochimica Acta*, **1988**, *1*, 301-304.

20. Nguyen, T.; Byrd, E. *J. of Coatings Technology*, **1987**, *59*, pp. 39-44.
21. Bretzlaff, R.S.; Sandlin, S.L. *Fourier Transform Infrared Spectroscopic Study of Thermal and Electrical Aging in Polyurethane*, SD-TR-87-04 The Aerospace Corporation, March 20, 1987.
22. Weber, D.; Fülöp, G.; Hummel, D.O. *Makromol. Chem., Macromol. Symp.* **1991**, *52*, pp. 151-160.

RECEIVED August 23, 1994

Chapter 11

Differential Scanning Calorimetry–Fourier Transform IR Spectroscopy and Thermogravimetric Analysis–Fourier Transform IR Spectroscopy To Differentiate Between Very Similar Polymer Materials

David J. Johnson[1], Philip J. Stout[2,3], Stephen L. Hill[2], and K. Krishnan[2]

[1]Applied Systems Inc., 200 Harry S. Truman Parkway, Annapolis, MD 21401
[2]Bio-Rad, Digilab Division, 237 Putnam Avenue, Cambridge, MA 02139

TGA/FT-IR and DSC/FT-IR were used to characterize an amine activated epoxy system. Cured and uncured systems were analyzed using these hyphenated techniques. The TGA/FT-IR study focused on measuring activator/resin ratios of cured epoxy systems for two different cure schedules. Evolved gases from cured epoxy systems were monitored and measured by FT-IR while the sample was being heated within the TGA oven. "Activator" and "Resin" specific gas profiles were integrated after calibrating with known standards to measure Activator/Resin ratios for unknown samples. During the same run, the shape and location of the specific gas and first derivative weight loss profiles were analyzed and compared to provide information relative to the thermal history of the epoxy system. Data from this research show that the utilization of the TGA/FT-IR technique allows the scientist to ensure proper mixing ratios while at the same time determine if the system was cured properly.

The same epoxy system in the uncured state was studied by DSC/FT-IR as a function of activator/resin ratio. Excess activator (over stoichiometric amounts) generated a single peak exotherm on the DSC trace while the activator deficient system produced a double peak exotherm. The DSC cell was placed on an FT-IR microscope stage. While the epoxy system was being heated, infrared spectra were continually collected by micro-reflection/absorption. Spectra from the two systems were compared and interpreted to explain on a molecular level the noted differences in the DSC curves. The FT-IR data confirmed that these detected differences were due to changes in the rate of cure and reaction mechanism.

[3]Current address: InoMet Inc., 2800 Patton Rd., Roseville, MN 55113

The study of polymeric materials by a combination of thermal analysis and infrared spectroscopy has become increasingly popular in today's analytical laboratory. Thermogravimetric analysis/Fourier transform infrared spectroscopy (TGA/FT-IR) has been used routinely to monitor the weight loss of a sample as a function of temperature while simultaneously monitoring the evolved gases. The coupling of an FT-IR to the TGA allows the chemist to assign the evolved gases to the detected weight losses and thereby correlate chemical information with the thermal event. Quantitative analysis may also be performed using the TGA/FT-IR technique even when more than one component of interest pyrolyzes during a single weight loss (1). This is accomplished by judiciously selecting the appropriate IR Specific Gas Profile (SGP) which isolates the pyrolysis product of the component you wish to measure. Although other gases may be evolving due to the pyrolysis of additional starting components, the absorbances due to these additional gases will not contribute to the generated SGP if the proper IR region is chosen.   After preparing a calibration with known standards, the generated profile is integrated. The calculated integrated absorbance is then related to the concentration of the starting component based on the calibration.

Differential scanning calorimetry/Fourier transform infrared spectroscopy (DSC/FT-IR) is utilized to monitor the chemical changes in the material itself as it is heated while simultaneously generating a DSC trace. The DSC measures the exothermic and endothermic responses of the samples as it is heated through various thermal transitions (2). Coupling of the FT-IR to the DSC allows the chemist to correlate structural changes in the material for each thermal event.

The DSC cell is placed on the stage of an infrared microscope. As the material is being heated, and while the DSC curve is being generated, infrared spectra are continually collected by micro-reflection/absorption. Spectra collected during a thermal transition may be compared and interpreted to explain, on a molecular level, the cause of this transition.

TGA/FT-IR and DSC/FT-IR were utilized in this research to characterize an amine activated epoxy resin system. The specific system under study was a near-monomeric diglycidyl ether of Bisphenol A (2-di-[4-(2,3-epoxy-1-propoxy)-1-phenyl]propane) with an epoxy equivalent weight of 173. The curing agent was composed of a mixture of 10% tertiary amine catalyst and 90% primary cycloaliphatic diamine. Cured and uncured systems were analyzed using these hyphenated techniques.

The TGA/FT-IR study focused on the pyrolysis products of the cured system to measure activator/resin ratios and during the same run, compare generated "Activator" and "Resin" specific gas and first derivative weight loss profiles to qualitatively determine the thermal history of the epoxy system. TGA/FT-IR results from epoxy systems subjected to varying cures were analyzed and compared.

The uncured epoxy system described above was analyzed by DSC/FT-IR as a function of activator/resin ratio. These samples were cured in the DSC cell. DSC traces were generated to characterize the curing process. Infrared spectra were collected simultaneously to monitor the cross-linking activity. The generated spectra for each mix were analyzed and compared to explain differences noted in the DSC curves.

EXPERIMENTAL

The TGA/FT-IR data were collected by using a Bio-Rad FTS 40 spectrometer coupled to a PL Thermal Sciences TGA 1000. A Bio-Rad TGA/IR interface bench was located between the spectrometer and TGA. The interface bench was equipped with a high temperature gas cell (where the evolved gases from the TGA experiment are analyzed by IR), an infrared detector (MCT), and two Watlows to heat and control the temperature of the gas cell and transfer line. The 1 mm ID transfer line was made of stainless steel with a thin silica lining.

The TGA/FT-IR system is totally integrated. The Bio-Rad 3200 data station collects all infrared spectra and chromatograms. All spectra were generated at 4 cm^{-1} resolution coadding 16 scans per scanset. The 3200 also simultaneously controls the heating of the furnace and collection of all thermograms. The weight loss data are transferred via an RS-232 cable from the PL Computer Controlled Interface (CCI) to the 3200 in real time.

All cured samples analyzed by TGA/FT-IR were heated from 30 to 600 °C at a rate of 20 degrees per minute. Sample sizes ranged between 6.5 to 12.5 milligrams. All samples were run under a nitrogen purge of approximately 30 ml/minute.

In the TGA/FT-IR research two cure schedules were employed for comparison. The first set of specimens was cured for four hours at 150 °C. This cure is referred to as a "zap" cure. The second set of samples was cured for eight hours at 57 °C, sixteen hours at 70 °C, followed by 24 hours at 110 °C, which is referred to as a "slow step" cure.

The DSC/FT-IR data were collected using a Bio-Rad FTS 40 spectrometer and UMA 300A infrared-transmitting microscope accessory. The sample under analysis was placed in a Mettler FP84 TA microscopy cell positioned on the microscope stage. To provide sufficient working distance to focus onto the sample within the DSC cup, a 15X Cassegrainian objective was utilized for both infrared analysis and viewing with visible light. The cell temperature was controlled by a Mettler FP80 central processor. Spectra were collected continuously during the heating of the sample by micro-reflection/absorption.

The uncured epoxy system was heated from 25 to 280 °C at a ramping rate of 10 degrees per minute. All samples were uncovered and open to atmosphere. Infrared spectra were collected at 8 cm^{-1} resolution coadding 128 scans per scanset.

The epoxy resin system under study was provided by Bacon Industries, Inc., 192 Pleasant Street, Watertown, MA 02172.

RESULTS AND DISCUSSION

TGA/FT-IR STUDY

The TGA/FT-IR research focused on the characterization of cured amine activated epoxy systems. This hyphenated technique was used to quantitate activator-resin ration for the cured system for two different cure schedules. The "zap" and "slow step" cures discussed in the experimental section are attractive processes for different reasons. The production line foreman would naturally prefer the "zap" cure since it would produce a high volume of parts in a short period of time. The polymer engineer favors the "slow step" cure since research has shown that this schedule produces the more thermodynamically stable product. The generated first derivative weight loss profiles and specific gas profiles were utilized to determine the cure schedule of these materials.

The TGA/FT-IR software allows for the generation of specific gas profiles (SGPs) which monitor the evolution of a gas absorbing within a specific frequency region as a function of time. An 877 - 904 cm^{-1} SGP was utilized to monitor and measure the evolution of 4-isopropenyl-methyl cyclohexene representing the pyrolysis product of the primary cycloaliphatic diamine. Likewise, an 1160 - 1185 cm^{-1} SGP was utilized to monitor and measure the evolution of 4,4-isopropylidenediphenol (Bisphenol A) corresponding to the pyrolysis product of the resin. The pyrolysis products were identified by searching the generated spectra against the Sadtler Vapor Phase Library. Search results identifying the two compounds mentioned above are shown in Figures 1 and 2. The unknown spectrum generated from the TGA/FT-IR run is the bottom spectrum in both figures. The top three "matches" are plotted in ascending order above the unknown.

The two generated SGPs were integrated to measure activator-resin ratios for both "zap" and "slow step" cured polymers. Figure 3 shows the 2 SGPs along with the Evolved Gas Profile which measures total IR absorbance as a function of time. By running a series of standards (cured using either the "zap" or "slow step" cure schedule), it was demonstrated that the integrated absorbance of these SGPs is proportional to the weight of the activator or resin. Figure 4 is a plot of the "activator" calibration which indicates that Beer's law is obeyed. The dark circles represent data generated from the "slow step" cured materials while the open squares represent data collected from the "zap" cured polymers.

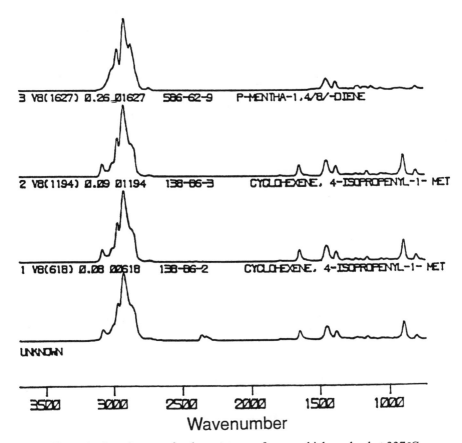

Figure 1. Search report for the spectrum of gases which evolved at 337 °C from a "slow step" cured epoxy sample (bottom trace). (Reproduced with permission from ref. 6. Copyright 1993 Elsevier Science Publishers B.V.)

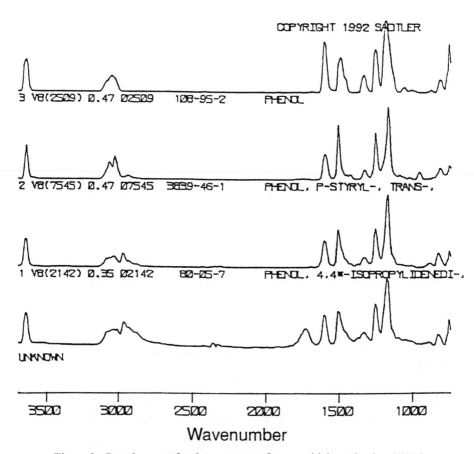

Figure 2. Search report for the spectrum of gases which evolved at 474 °C from a "slow step" cured epoxy sample (bottom trace). (Reproduced with permission from ref. 6. Copyright 1993 Elsevier Science Publishers B.V.)

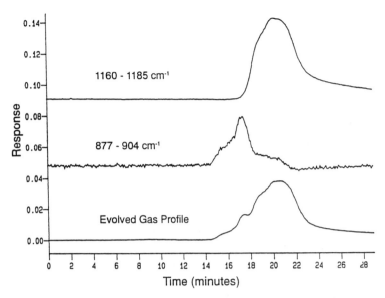

Figure 3. The Specific Gas Profiles for "resin" (1160 - 1185 cm^{-1}) and "activator (877 - 904 cm^{-1}) and the total Evolved Gas Profile generated during a TGA/FT-IR run of a "slow step" cured epoxy sample. (Reproduced with permission from ref. 6. Copyright 1993 Elsevier Science Publishers B.V.)

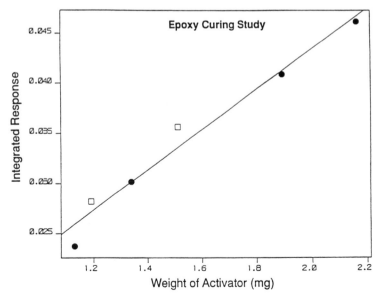

Figure 4. A plot of integrated absorbance of the 875 - 904 cm^{-1} Specific Gas Profile vs. weight of activator from known standards of "slow step" (dark circles) and "zap" cured (open squares) epoxy samples. The least-squares fit shown is for the complete data set. (Reproduced with permission from ref. 6. Copyright 1993 Elsevier Science Publishers B.V.)

When examining Figure 3, note that the evolution of the activator and resin do overlap in time. As a result, weight loss data alone could not be used to measure the individual contributions of each component. Since each pyrolysis product exhibits a unique infrared absorbance spectrum, their evolutions may be separated and measured through judicious selection of SGPs.

It should be noted that a slower heating rate may have been utilized to yield better separation. A 20 °C per minute ramping rate was used in this research to maximize the rate of gas evolution and thereby increase the signal of the generated SGPs. Even if the ramping was decreased, without confirmation from the FT-IR, complete separation could not be guaranteed.

Two "Activator" SGPs generated from a "slow step" and a "zap" cured system are plotted for comparison in Figure 5. Note that the shape and location on the time scale of the profiles vary considerably. This plot shows that the majority of the activator pyrolyzes in the "zap" cured material at lower temperatures relative to the "slow step" cured system. This is to be expected since the activator is not as extensively crosslinked to the resin during a "zap" cure and as a result more easily "fragmented" from the system upon heating. Theoretically, a series of profiles may be generated as a function of cure schedule. A profile of an "unknown" material may be generated and compared or "searched" against this series of known cures to

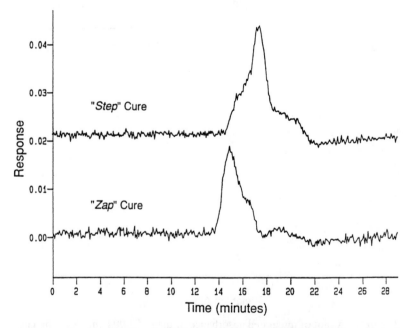

Figure 5. The Specific Gas Profiles for "activator" (877 - 904 cm^{-1}) generated from a "slow step" and "zap" cured epoxy sample, plotted on a common time axis. (Reproduced with permission from ref. 6. Copyright 1993 Elsevier Science Publishers B.V.)

estimate prior thermal history. It is important to note that although the location and position of the profiles change as a function of cure, the area of these profiles seen to remain proportional to the activator (or resin) regardless of cure schedule. It should also be noted that the generated first derivatives also vary as a function of cure. The first derivative of an "unknown" may also be "searched" for the same purpose.

DSC/FT-IR STUDY

Mirabella and Koberstein have previously shown the benefit of DSC/FT-IR for polymer characterization (3,4). In this work, the same epoxy system described above in the uncured state was analyzed by DSC/FT-IR. Thin films of uncured amine-activated epoxies were placed in the sample pan of the FP84 and heated from 25 to 280 °C at 10 °C per minute. Changes in the structure of the epoxy as a function of temperature were recorded simultaneously by infrared spectroscopy. The sample was relatively transmissive to infrared radiation. The beam transmitted down through the sample, reflected off the aluminum cup, and passed back up through the material. This type of analysis is called reflection/absorption spectroscopy. A "well behaved" absorbance spectrum was generated directly without any need for correction. To produce a sufficient signal on the DSC, the bulk of the sample had to be placed on the reference side.

The curing of an epoxy is an exothermic reaction. As the activator-resin ratio changes the generated DSC peak varies considerably. Figure 6 compares two traces of the same epoxy system of different mixes. Note that the ordinate is reversed in

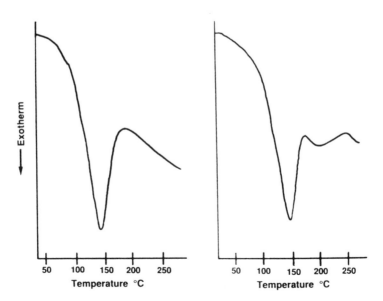

Figure 6. Generated DSC traces during cure of 100 parts resin to 35 parts (left) and 17 parts (right) activator systems. (Reproduced with permission from ref. 7. Copyright 1992 Elsevier Science Publishers B.V.)

both traces since the curves were generated with the bulk sample located in the reference pan. The two mixes were 100 parts resin to 35 parts activator (41% over the stoichiometric primary and secondary amine concentration) and 100 parts resin to 17 parts activator (31% under the stoichiometric level). Note that when less activator is used the exotherm changes from a single peak to a double peak exotherm. This indicates a change in reaction mechanism. By coupling the FT-IR to the DSC these changes were studied and differentiated.

The general curing reaction mechanism of this system is relatively straight forward. The cure initially involves the reaction of the cycloaliphatic primary amine activator (4-amino- 4-trimethyl-cyclohexanemethaneamine) with the epoxide group of the diglycidyl ether of Bisphenol A to produce a secondary amine. This secondary amine further reacts with additional epoxide groups to produce a tertiary amine. Further reaction, catalyzed by water, hydroxyl, and tertiary amine concentration, continues the crosslinking activity.

A series of infrared spectra are generated during the curing process as shown in Figure 7. Relative band heights are measured to track the curing of each sample. These curing profiles may be compared as a function of activator-resin ratio. Results as shown in Figures 8 and 9 indicate that the mixing ratio influences the rate of reaction and also changes the reaction mechanism.

Figure 8 monitors the changes in the relative peak heights of the 3030 cm^{-1} band to the 3048 cm^{-1} band for the two mixes. Both bands are due to the aromatic C-H stretching of the polymer. Comparing the slopes of these plots indicates that the rate of the reaction for the over-stoichiometric mix is faster than the 17 parts activator system.

Examination of the plots in Figure 9 indicates that varying the activator-resin ratio may also change the reaction mechanism of the cure. Figure 9 monitors the changes in the relative peak heights of the 3467 cm^{-1} band (hydroxyl group) divided by the 1600 cm^{-1} band (aromatic C-C skeletal stretch) for the two mixes. These plots may be used to monitor the hydroxyl formation during cure. The initial higher rate of hydroxyl formation in the mixture having the higher activator level again indicates that the over-stoichiometric mix increases the reaction rate. It is interesting to note the decrease in the rate of hydroxyl formation in the over-stoichiometric mix at approximately 210 °C. This may be due to an etherification/homopolymerization reaction, consuming hydroxy groups, which competes with the secondary amine for epoxy sites (5).

Close examination of the individual spectra indicate that the cycloaliphatic diamine shows a strong preference for the primary amine reaction. With an excess of activator the primary amine reaction dominates resulting in a single exotherm. In the under-stoichiometric mix, excess epoxide groups remain after the majority of the primary amine has been consumed. The less reactive secondary amine apparently

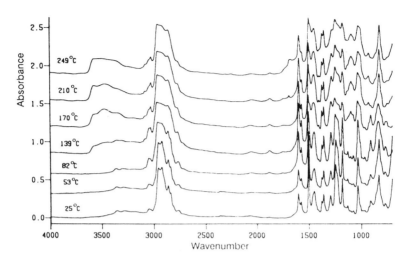

Figure 7. A series of IR spectra generated at various temperatures during the heating of an epoxy sample containing 35 parts activator. (Reproduced with permission from ref. 7. Copyright 1992 Elsevier Science Publishers B.V.)

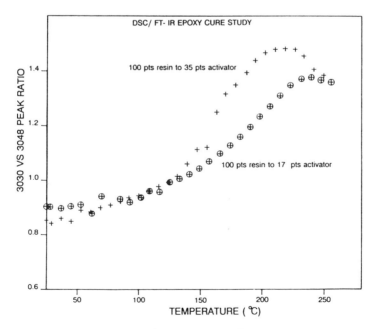

Figure 8. A plot of 3030 cm⁻¹ versus 3048 cm⁻¹ peak ratio as a function of temperature for both mixes. (Reproduced with permission from ref. 7. Copyright 1992 Elsevier Science Publishers B.V.)

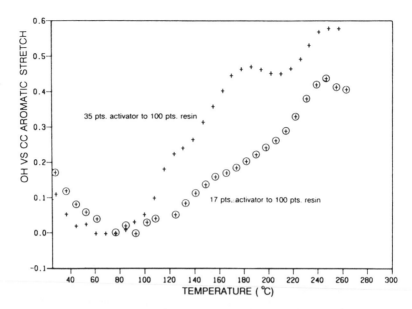

Figure 9. A plot of 3467 cm⁻¹ (hydroxyl group) versus 1600 cm⁻¹ (aromatic group) as a function of temperature for both mixes. (Reproduced with permission from ref. 7. Copyright 1992 Elsevier Science Publishers B.V.)

then plays a major role in the formation of the second DSC peak.

CONCLUSIONS

Results indicate that the study of polymeric materials by a combination of thermal analysis and infrared spectroscopy provides the chemist with a wealth of information. The infrared data may be used to provide needed chemical information about a polymer as it is heated through various thermal transitions. In the case of TGA/FT-IR, important information is gained relative to the evolved gases which are responsible for the detected weight losses. Specific Gas Profiles may be created to monitor the evolution of pyrolysis products of specific starting components. The proper selection of these SGPs will allow the chemist to measure the amount of an individual starting component even if other materials are .pyrolyzing during the same weight loss. Results from this research suggest that starting activator and resin concentrations may be measured using the TGA/FT-IR technique regardless of cure schedule.

DSC/FT-IR allows the chemist to monitor the chemical changes of the material itself as it is heated. The coupling of an FT-IR to a TGA or DSC provides the chemical information necessary for interpretation of the thermal event. The generation of DSC traces during the cure of an amine activated epoxy system as a function of activator-resin ratio indicates that the mix ratio will influence the rate of cure and/or alter the reaction mechanism. The examination of the infrared spectra of the material as it was heated varifies that both the rate and mechanism of cure vary as a function of mix ratio. The additional molecular information provided by the FT-IR significantly aided in the interpretation of the generated DSC traces.

LITERATURE CITED

1. Johnson, D.J.; Compton, D.A.C , *Amer. Lab.* January **1991**, pp. 37 - 43.

2. *Thermal Characterization of Polymeric Materials*; Turi, E., Ed.; Academic: Orlando, FL, 1981

3. Mirabella, F.M. *Appl. Spectrosc.* **1986**, 40, pp. 417 - 420.

4. Koberstein, J.T.; Gancarz, I.; Clarke, T.C. *J. Polym. Sci., Polym. Phys. Ed.* **1986**, 24, pp. 2487 - 2498.

5. Riccardi, C.C.; Williams, R.J.J. *J. Appl. Polym. Sci.* **1986** 32, pp. 3445 - 3456.

6. Johnson, D.J.; Compton, D.A.C.; Cass, R.S.; Canale, P.L. *Thermochimica Acta*, 000, **1993**, pp. 1-16.

7. Johnson, D.J.; Compton, D.A.C.; Canale, P.L. *Thermochim. Acta* **1992**, 195, pp. 5-20.

RECEIVED July 26, 1994

Chapter 12

Polymer Structure Determination Using Simultaneous Small- and Wide-Angle X-ray Scattering and Differential Scanning Calorimetry

Anthony J. Ryan[1,2], Steven Naylor[1], Bernd Komanschek[1], Wim Bras[2,3], Geoffrey R. Mant[2], and Gareth E. Derbyshire[2]

[1]Manchester Materials Science Center, University of Manchester Institute of Science and Technology, Manchester M1 7HS, England
[2]DRAL, Daresbury Laboratory, Warrington WA4 4AD, England
[3]Nederlandse Organisatie voor Wetenschappelijk Onderzoek, P.O. Box 93138, 2509 AC, Haague, The Netherlands

Experimental technique for the characterisation of the thermal-morphological properties of materials by X-ray scattering have been developed at Daresbury. Many thermal events, for example melting endotherms, are signals of phase transitions and thus changes in morphology. Collection of DSC and X-ray patterns simultaneously aids interpretation of the thermal behaviour. In many systems that morphology covers size-scales from the atomic to the microscopic, *i.e.* Å to μm. There are obvious advantages in collecting both the wide angle (sizes from 1 to 20 Å) and small angle (sizes from 20 to 1000 Å) X-ray patterns simultaneously to unambiguously characterise such thermal events. The new SAXS/WAXS technique makes this possible with a time resolution of 0.1 second which allows heating rates up to 120 °C min^{-1}. Simultaneous SAXS/DSC and SAXS/WAXS techniques are shown to provide an unambiguous method to follow the structural changes taking place during the programmed heating of a range of multiphase polymeric materials. The principles of the experiments are illustrated with specific examples of polyethylene terepthalate, high density polyethylene and block copolyurethanes.

The experimental technique of DSC or DTA is often used in the thermal characterisation of structure-property-relations in polymers [1]. The phenomena investigated, such as melting and glass transitions in semi-crystalline polymers or the glass transitions of blends, are associated with strong morphological features. Much of the knowledge concerning the crystallisation of polymers comes from the application of programmed heating and cooling, and isothermal crystallisation studies by DSC combined with post-mortem assessment of morphology by either X-ray diffraction or microscopy. Similarly, the existence of phase separation in polymer blends [2] or microphase separation in block copolymers [3] is often assessed by DSC with confirmation sought by scattering, microscopy and dynamic mechanical thermal analysis. The kinetics of

0097–6156/94/0581–0162$08.00/0

(micro)phase separation can be followed by DSC but the thermal response is weak and these experiments are often semi-quantitative [4].

There is a reciprocity between the scattering angle and the size scale of the structure probed. The technique of wide angle X-ray diffraction (WAXS) probes atomic distances (1 - 20 Å) and may be used to solve the crystal structure (establish the unit cell and atomic positions) if the full diffraction pattern of a single crystal or fibre is available [5]. WAXS is based on the scattering of radiation by electrons where interference effects are correlated at atomic dimensions and it may be used to calculate atomic positions. Due to the polycrystalline nature of most polymers it is more common to obtain the radially symmetric 1-dimensional, 1-D, powder diffraction pattern and, wherever possible, index the structure from this. For example, the two crystalline forms of polyethylene can be readily distinguished from their powder diffraction patterns [6]. Small angle X-ray scattering (SAXS) is a well established technique for studying the morphology of multiphase polymers [6] and probes larger length scales (50 - 1000 Å). In SAXS, X-rays are scattered by electrons in regions with different electron densities and the correlations are longer ranged. It is often used in tandem with DSC to study polymer crystallisation and microphase separation in block copolymers. Information is obtained in the form of a scattering pattern; as with WAXS unoriented materials have radially symmetrical 1D patterns which can be analysed using Bragg's law leading to information on the structural features with size-scales from 50 - 1000 Å. For liquid crystalline and semi-crystalline polymers this corresponds to the crystallite size, for block copolymers the unit cell. In some cases, where the data is of a very high statistical quality (*i.e.* a high signal to noise ratio), correlation function analyses can yield further spatial information such as the thickness of the interface between microphases in a block copolymer or between the crystalline and amorphous regions of a semicrystalline polymer [6].

Conventional SAXS and WAXS experiments, that is those utilising sealed-tube or rotating anode X-ray generators, are limited to stable materials due to the long times (hours) required to obtain patterns of sufficient statistical quality or spatial resolution. Patterns may be taken as a function of temperature but generally this is not done due to the difficulties of furnace design. A good review of what has been done with conventional X-ray sources is given by Charlesly [7]. Furnaces must generally contain windows that will support molten polymers and be transparent to X-rays, suitable materials are mica and cross linked polyimide. The major problem is with leakage, after long times at high temperatures, and subsequent camera contamination. To obtain statistically significant SAXS and WAXS patterns at heating rates used in DTA or DSC experiments (time resolution of less than one minute and preferably less than one second) the high flux of synchrotron radiation and fast, position-sensitive, electronic detectors must be used. Synchrotron radiation is produced at special facilities (such as SSRL at Stanford, California; NSLS, New York; The Photon Factory, Osaka; DESY, Hamburg and the SRS at Daresbury, UK) by maintaining electrons in a relativistic orbit, the radiation is taken off at ports which are at a tangent to the electron trajectory. The radiation diverges slightly but can be easily monochromated and focused using X-ray mirrors and monochromators. It is thus possible to obtain low-divergence, point-collimated, monochromatic X-ray sources. There are around twenty suitable synchrotrons world-wide each containing up to sixty beamlines. [8] New sources are being constructed in Grenoble (ESRF), Japan (Spring-8) and Chicago (APS).

The combined techniques of DSC/SAXS and DSC/WAXS were pioneered by Koberstein and Russell [9] using synchrotron radiation (SSRL) as the X-ray source and a Mettler FP-85 DSC which had been modified to allow transmission of radiation beams. Experiments on low density polyethylene, a well characterised material, were performed in order to validate the technique [9]. A large contribution to the understanding of the complex behaviour of thermoplastic copolyurethanes has been made by the detailed study using DSC/FTIR [10], DSC/SAXS [11,12] and DSC/WAXS [13] by Koberstein and co-workers. The morphologies formed and the kinetics of their formation have been deduced by separate DSC/SAXS and DSC/WAXS experiments. The inevitable conclusion of Koberstein's work is that combined SAXS/WAXS is a step toward minimising experimental effort and realising unambiguous results by removing the need for repeated experiments on different samples.

Ungar and Feijoo [14] have also used DSC/SAXS and DSC/WAXS (using SRS as the radiation source and a modified Mettler FP85) to study chain folding in the crystallisation of high molecular weight, monodisperse, linear alkanes and the morphological transitions of side chain liquid crystalline polymers. The need for simultaneous SAXS/WAXS is also borne out by this work.

The polymer beam-line at DESY has been used for time and temperature resolved studies of polymers notably the work of the Hamburg group on polyesters [15] and polyolefins [16]. The SAXS camera is equipped with a Gabriel-type detector and provides for good spatial and time resolution. At NSLS Chu and co-workers have recently been prolific with their studies of polyurethanes [17-19] and ionomers [20,21] by synchrotron SAXS. The SAXS geometry is an under focused Kratky camera with a linear detector and is equipped with a fast T-jump facility. There is also a pin-hole collimated SAXS camera at NSLS and this has also been used for time-resolved SAXS from polymer systems [22].

An ideal simultaneous SAXS/WAXS experiment would provide spatial information over size-scales of 1000 Å to 1 Å as illustrated in Figure 1. The SAXS experiments allow calculation of the lamella spacing from the peak maximum and calculation of the invariant. The crystal structure may be deduced from the positions of the peaks in the WAXS pattern and the degree of crystallinity could be calculated from the ratios of intensities. The synergy of information available from time-resolved SAXS/WAXS has attracted much interest and three such instruments have been constructed.

Advances have been made at DESY in Germany, Zachmann [23] made the first attempt at combining time-resolved WAXS and SAXS to study PET and its related liquid crystalline polymers. Due to the experimental arrangement used (vidicon and linear detectors) the data had a poor signal to noise ratio at DSC heating rates. This pioneering work cannot be ignored however and despite the limitations of the signal to noise ratio in calculating phase compositions good spatial information was obtained. The most recent manifestation of SAXS/WAXS from DESY [24] resorted to repeating the WAXS part of the experiment using a linear Gabriel detector in order to improve the counting statistics and spatial resolution.

Great strides have been made by Chu and co-workers at the NSLS where a SAXS/WAXS camera has recently been constructed and used to study [215] the melting and crystallisation of polyolefin blends and the melting and recrystallisation of PEEK [26,27]. Although their camera had good spatial resolution in both the small and wide angle regimes the angular range for WAXS was limited by the use of a linear detector to $\approx 10°$ of arc and the SAXS camera was compromised by under focused Kratky optics.

At Daresbury we have recently constructed a SAXS/WAXS camera on a high intensity, highly-collimated beam line optimised for isotropic scatterers and equipped with a quadrant detector for SAXS and a curved detector for WAXS that covered > 70° of arc [28,29]. The problems associated with the two previously reported set-ups [23,25]were mainly alleviated. The experimental set-up has been described in detail [29] and its context reviewed [30]. The science addressed using this instrument includes the study of changes in crystal parameters during the melting and recrystallisation of HDPE [31], a crystal-crystal transformation in polypropylene [32], problems in the microphase separation of polyurethanes [33], structural changes during gelation of starch solutions (cooking) [34] and some of these developments will be reviewed herein. Most recently the SAXS camera was equipped with a DSC as the sample holder and furnace and this will be described for the first time here.

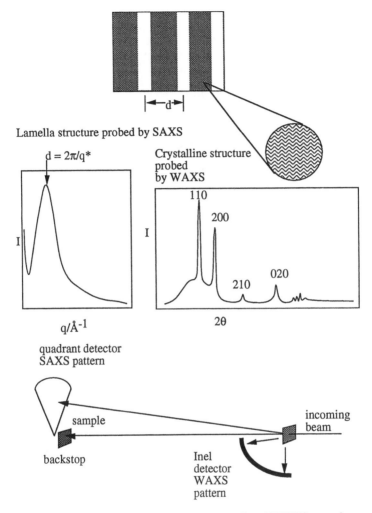

Lamella structure probed by SAXS

$d = 2\pi/q^*$

Crystalline structure probed by WAXS

I

I

q/Å^{-1}

2θ

quadrant detector
SAXS pattern

sample

incoming beam

backstop

Inel detector WAXS pattern

Figure 1. Spatial information available from SAXS and WAXS experiments on semicrystalline polymers. The SAXS experiments allow calculation of the lamella spacing from the peak maximum. The crystal structure may be deduced from the positions of the peaks in the WAXS pattern. The camera geometry is also included.

Experimental

Simultaneous SAXS/WAXS measurements were made on beam line 8.2 of the SRS at the SERC Daresbury Laboratory, Warrington, UK. The details of the storage ring, radiation and camera geometry and data collection electronics have been given in detail elsewhere [35]. The camera is equipped with a quadrant detector (SAXS) located 3.5 m from the sample position and a curved knife-edge detector (WAXS) that covers 120 ° of arc at a radius of 0.2 m. A vacuum chamber is placed between the sample and detectors in order to reduce air scattering and absorption. Both the exit window of the beam-line and the entrance window of the vacuum chamber are made from 15 μm mica, the exit windows of the vacuum chambers are made from 15 μm mica and 10 μm Kapton film for the WAXS and SAXS detectors respectively. The WAXS detector has a spatial resolution of 50 μm and can handle up to ~100 000 counts s^{-1} ; only 90° of arc are active in these experiments the rest of the detector being shielded with lead. A beam stop is mounted just before the SAXS exit window to prevent the direct beam from hitting the SAXS detector which measures intensity in the radial direction (over an opening angle of 70 ° and an active length of 0.2 m) and is only suitable for isomorphous scatterers. It has an advantage over single-wire detectors in that the active area increases radially improving the signal to noise ratio at larger scattering angles. The spatial resolution of the SAXS detector is 400 μm and it can handle up to ~500 000 counts s^{-1}.

The specimens for SAXS/WAXS and SAXS/DSC were prepared by placing a disks ≈ 0.5 mm thick and ≈ 6 mm diameter, cut from premoulded sheets, in a cell comprising a Du Pont DSC pan fitted with windows (≈4 mm diameter) made from 5 μm thick mica. The modified DSC pans are described elsewhere [11,30].

Loaded, sealed pans were glued to a J-type thermocouple and placed in a spring loaded holder in a modified Linkam TMH600 hot-stage mounted on the optical bench for SAXS/WAXS. The silver heating block of the hot-stage contains a 4 x 1 mm conical hole which allows the incident X-rays to pass through unhindered. A nominal heating rate of 10 °C min^{-1} was used for SAXS/WAXS and due to the nature of the furnace there was a small temperature gradient across the sample chamber which required direct measurement of temperature at the sample position over and above the temperature monitoring and control of the furnace.

The Linkam DSC is of the single specimen design and is described in detail elsewhere [36]. The cell comprises a silver enclosure around a thermocouple plate, the plate has a 3 x 0.5 mm slot and the sample is held in contact with the plate by a low thermal mass spring. A reference (calibration) sample of the same thermal mass was first subjected to the temperature ramp and the thermal response of the neutral system recorded. The sample was then run and its thermal response recorded. The differential response was subsequently calculated from the electronic recordings, the single pan technique relies on the accuracy and reproducibility of the temperature control system and is comparable to that of a conventional, two-pan, heat-flux DSC (for example a Du Pont 990) [1].

The scattering pattern from an oriented specimen of wet collagen (rat-tail tendon) was used to calibrate the SAXS detector and HDPE, aluminium and an NBS silicon standard were used to calibrate the WAXS detector. A parallel plate ionisation detector placed before and after the sample cell recorded the incident and transmitted intensities. The experimental data were corrected for background scattering (subtraction of the scattering from the camera, hot stage and an empty cell), sample thickness and transmission, and the positional nonlinearity of the detectors.

Results

I   SAXS/DSC of polyethylene terepthalate (PET).   A peak or a shoulder in $I(q)$ versus q plot gives a general indication of the presence of a periodic structure in the system. The most common practice for determining the periodicity is to use Bragg's law in the calculation of a domain spacing, d, from the location of the peak maximum, q_{max}, in an intensity versus scattering vector plot .

$$d = \lambda / 2sin \ \theta_{max} = 2\pi / q_{max} \qquad \{1\}$$

There is a weak peak in the bare intensity data for PET at 228 °C, Figure 2, making unambiguous determination of q_{max} difficult . For the semi crystalline polymers in the present study, more precise information regarding micro structural periodicity is obtained if the morphology is assumed to be globally isotropic but locally lamella [5]. The data can then be analysed to give a one-dimensional Bragg spacing, d_l, by applying the Lorentz correction, q^2, to the observed scattered intensity $I(q)$. Figure 2

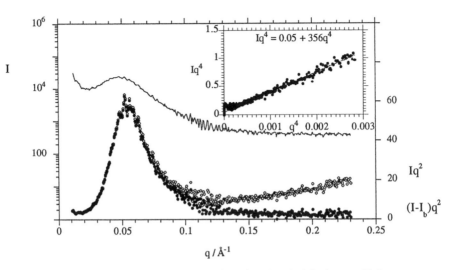

Figure 2. SAXS patterns of PET at 200°C. The line (—) is the raw $I(q)$ versus q data and has a weak peak at q_{max} and a strong upturn at low q. The effects of diffuse scattering from the thermal background at high q are obvious in the uncorrected data, (o) $I(q)q^2$ versus q, and their corrected values, (●) $(I(q)-I_b)q^2$ versus q. The inset is the Porod plot used to calculate the thermal background from the slope of the linear portion in Iq^4 versus q^4.

includes a plot of $I(q)$ and $I(q)\ q^2$ versus q for PET at 228 °C. For materials with sharp phase boundaries, Porod's Law [37] predicts a fall off of q^{-4} in scattered intensity at large angles.

$$\lim_{q\to\infty} I(q) = (K_p\ /\ q^4) + I_b \qquad \{2\}$$

The quantity I_b arises from density fluctuations and K_p is the Porod constant. Note the strong positive deviation in the Porod plot ($I(q)q^4$ versus q^4) illustrated as an inset to Figure 2. Positive deviations from Porod's law are caused by thermal density fluctuations [6]. A regression analysis of the linear part of the curve gives values of K_p = 0.05 and I_b = 356. I_b must be subtracted from the raw intensity data before both d_1 and the invariant are calculated.

The filled symbols in symbols in Figure 2 are values of $(I- I_b)q^2$ versus q and the artificial upturn of $I(q)\ q^2$ at high q has been corrected (open symbols). For the calculation of the Bragg spacing, the maximum in $(I- I_b)q^2$ versus q is taken as q^*, so that $d_1 = 2\pi/q^*$. The peak value from $I(q)$ versus q gives a d spacing of ≈ 134 Å whereas the peak value from $(I- I_b)q^2$ versus q gives a d_1 spacing of 121 Å The fully-corrected d_1 spacing will be used in the discussion.

Time-resolved SAXS data were collected during heating and cooling DSC experiments with a heating rate of 20 °C min^{-1} and a time resolution of 6 s. Figure 3 has a three dimensional plot of Lorentz corrected intensity, $I(q)q^2$, versus scattering vector, q, versus temperature, T, in the temperature interval 40 to 390 °C for quenched PET. The corresponding iso-intensity contours are shown in the contour plot. Initially the material is completely amorphous and there is no small angle scattering, at ≈ 130 °C there is the first appearance of a small angle peak due to the crystals with $q^* = 0.05$ Å^{-1}, this peak grows in intensity and shifts to lower q as the melting point is reached at ≈ 265 °C. The shift in d_1 is most easily observed in the contour plot.

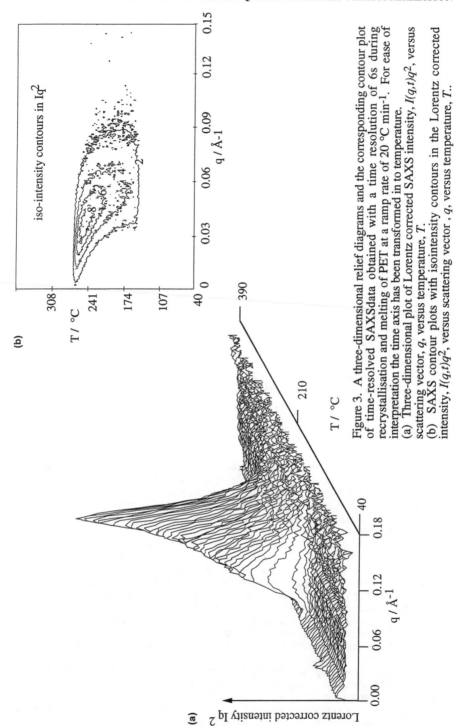

Figure 3. A three-dimensional relief diagrams and the corresponding contour plot of time-resolved SAXSdata obtained with a time resolution of 6s during recrystallisation and melting of PET at a ramp rate of 20 °C min^{-1}. For ease of interpretation the time axis has been transformed in to temperature.
(a) Three-dimensional plot of Lorentz corrected SAXS intensity, $I(q,t)q^2$, versus scattering vector, q, versus temperature, T.
(b) SAXS contour plots with isointensity contours in the Lorentz corrected intensity, $I(q,t)q^2$, versus scattering vector , q, versus temperature, T..

The invariant, Q, which is linear in the electron density difference $<\eta^2>$, between the crystalline and amorphous phases and quadratic in the volume fraction of crystals, ϕ, may be obtained from the integral in the equation

$$Q = \phi(1-\phi)\langle\eta^2\rangle = \frac{1}{2\pi i_e}\int\limits_0^\infty I(q)q^2 dq \qquad \{3\}.$$

where ϕ is the volume fraction of crystals and $\langle\eta^2\rangle$ the square of the electron density difference between the crystalline and amorphous phases and i_e is the Thompson scattering factor. If $\langle\eta^2\rangle$ for the system is known then the quadratic in ϕ may be readily solved. The absolute value of the invariant requires absolute intensity measurements, thermal background subtraction, and extrapolation to $q = 0$ and ∞ which is computationally difficult to achieve. The major contribution to the experimental invariant can be used to characterise structure development, as well as the degree of microphase separation, this is readily assessed by , for example, Simpson's rule integration of the the $I(q)$ q^2 versus q curve between experimental limits [9, 12, 13]. A relative invariant, Q', has been calculated from the area under the $(I(q)-I_b)q^2$ versus q curve between the first reliable data point, $q= 0.01$ Å^{-1}, and the region in which $(I(q)-I_b)q^2$ becomes constant, that is at $q= 0.20$ Å^{-1} according to

$$Q' = \int\limits_{0.01}^{0.20}(I(q)-I_b)q^2 dq \qquad \{4\}$$

Due to the relative nature of the intensity measurement the value of Q' is also only relative with arbitrary units.

The experimentally derived Q' and its differential with respect to time are shown in Figure 4 along with the simultaneously obtained DSC curve. The T_g is obviously not observed in the SAXS patterns but the temperature correlation of the growth in the invariant associated with crystallisation and exotherm from the DSC is readily observed. The maximum in dQ'/dT and T_c from DSC coincide within the time resolution of the SAXS experiment. The invariant continues to grow as heating continues and there are two contributions to this increase in Q'. $\langle\eta^2\rangle$ increases with T as the thermal expansion coefficients of the crystalline and amorphous regions differ, however, the dominant reasons for the increase in Q' is the continuous increase in the degree of crystallinity as lamella thickening and reorganisation occurs. The invariant passes through a maximum as 50% crystallinity is reached at ≈ 250 °C and Q' then falls dramatically as the polymer melts. The minimum in dQ'/dT and T_m from DSC coincide within the time resolution of the SAXS experiment.

II   SAXS/WAXS of high density polyethylene (HDPE). Typical data obtained during the melting and crystallisation of HDPE are presented in Figure 5, the patterns were collected with a time resolution of 10 s between 100 and 160 °C. Three-dimensional relief diagrams and the corresponding contour plots of time-resolved, Lorentz corrected SAXS and time-resolved WAXS data obtained at 10 s per pattern during melting and recrystallisation of high density polyethylene at a ramp rate of 6 °C min^{-1}. For ease of interpretation the time axis has been transformed into temperature. The SAXS patterns show a strong peak in $I(q)q^2$ (at q^*) which increases in intensity and then collapses toward the beam stop as the polymer melts, the WAXS data show strong peaks (which have been indexed to the orthorhombic structure) which drop continuously in intensity as the material melts. These patterns are consistent with individual lamellae melting in a first order thermodynamic process, the degree of crystallinity is falling (WAXS intensities reduce) and the average correlation length increasing (peak moves to lower q^*). The recrystallisation processes illustrates the

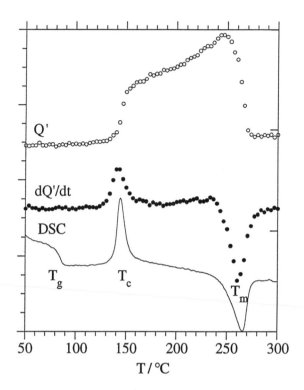

Figure 4. The relative invariant, Q', (o) and its differential with respect to temperature dQ'/dT (●) compared to the DSC curve(—) for PET. The thermal transitions are marked.

reversibility of the whole process, a weak SAXS peak appears at low q and grows through a maximum in intensity to a constant q^* and the WAXS peaks just grow to a maximum. These recrystallisation patterns are consistent with sporadic nucleation of lamellae that grow laterally and thicken.

The SAXS patterns for HDPE have been analysed in a similar manner to those for PET and the values of d_I and Q' calculated after subtraction of the Porod background. The compression moulded HDPE material has an initial d_I of 273 Å and the recrystallised material has d_I of 299 Å. The value of Q' calculated from the SAXS patterns is given in Figure 6 and the maximum in Q' associated with 50% crystallinity will be used to calibrate the integrated WAXS intensities.

The WAXS patterns in Figures 5c and 5d have been indexed according to literature [6] and the increased lattice expansion (shift in 2θ and line broadening)of the 200 compared to the 110 is well established [6]. There are two types of measurements of the degree of crystallinity by WAXS. External comparison methods are those involving the comparison of an experimental WAXS pattern to those of wholly crystalline and wholly amorphous standards. Internal comparison methods are those which use the integrated intensities of the pattern associated with the amorphous and crystalline features. To first approximation, the degree of crystallinity can be obtained by assuming that the total scattering within a certain region of reciprocal space is independent of the state of aggregation of the polymer. The degree of crystallinity

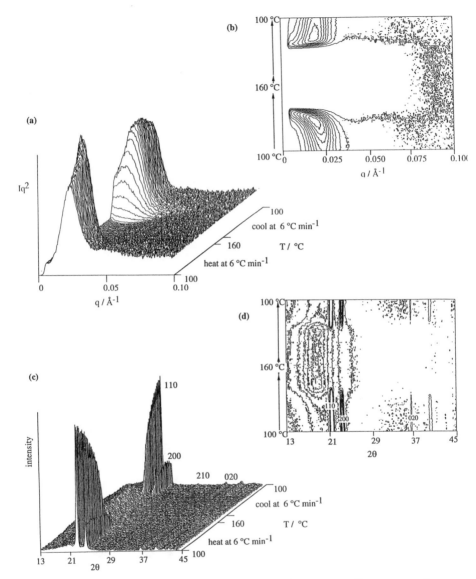

Figure 5. Three-dimensional relief diagrams and the corresponding contour plots of time-resolved SAXS and WAXS data obtained with a time resolution of 10s during melting and recrystallisation of high density polyethylene at a ramp rate of 6 °C min⁻¹. For ease of interpretation the time axis has been transformed in to temperature.
(a) Three-dimensional plot of Lorentz corrected SAXS intensity, $I(q,t)q^2$, versus scattering vector, q, versus temperature, T.
(b) SAXS contour plots with isointensity contours in the Lorentz corrected intensity, $I(q,t)q^2$, versus scattering vector , q, versus temperature, T..
(c) Three-dimensional plot of WAXS intensity, $I(2\theta,t)$, versus scattering vector, q, versus temperature, T.
(d) WAXS contour plots with isointensity contours in the scattered intensity, $I(2\theta,t)$, versus the scattering angle 2θ, versus temperature.

(mass fraction) may then be found from

$$X_c = I_c / I_t \qquad \{5\}$$

where I_c is the contribution of the crystalline component to the total scattering $I_t = I_c + I_a$. An example of this simple type of calculation for polyethylene is given in the introductory polymer text book of Young and Lovell [37] as being the ratio of the areas under the 110 and 200 peaks, viz.

$$X_c = A_c / (A_c + A_a) \qquad \{6\}$$

where A_a is the area under the amorphous and A_c is the area remaining under the crystalline peaks. The area under the strongest reflection, the 110, has been calculated by integration between the limits $20.9 < 2\theta < 21.9$ as a first order approximation to the degree of crystallinity. The integrated intensity at $T >> T_m$ gives the value of I_a and the integrated intensity at $T < T_m$ gives the value of $I_t = I_c + I_a$ allowing I_c to be calculated. Without performing numerical analysis on the whole pattern or in the absence of any method of calibration, the integrated intensity of a single reflection may only be used to get semi-quantitative information on the degree of crystallinity. The magnitude of I_c

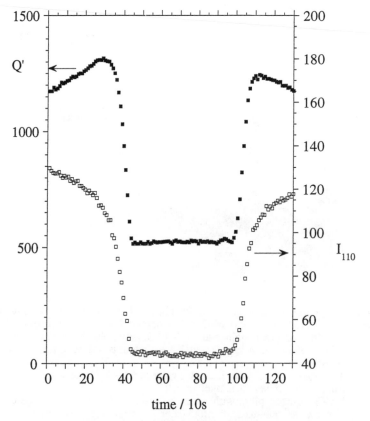

Figure 6. Time dependence of the SAXS invariant, Q' (■), and integrated WAXS intensity I_{110} (□), during melting and recrystallisation of HDPE.

when $X_c = 0$ $(T >> T_m)$ is known and the magnitude of I_c when X_c is a maximum is known but unless there is external calibration of the maximum in X_c then the measurements are arbitrary and qualitative.

The relative invariant, Q', from SAXS and the integrated intensity from WAXS, I_{110}, are plotted as a function of time (during a heating and cooling experiment) in Figure 6. In the first 600 s of heating the value (in arbitrary units) of I_{110} falls continuously from 130 ± 3 to a constant 50 ± 2 whereas Q' initially increases from 1250 ± 10 to a maximum 1330 ± 10 before falling to a constant value of 522 ± 7. The constant values at 600 s ($T = 160$ °C) are those of the amorphous molten polymer. The maximum in Q' is due to the relationship

$$Q' = \phi(1 - \phi)\langle \eta^2 \rangle \qquad \{7\}$$

which is linear in the electron density difference $\langle \eta^2 \rangle$ and quadratic in the volume fraction of crystals, ϕ [3, 15, 27]. If we assume the two-phase model [14] applies and $\langle \eta^2 \rangle$ is a constant, then the invariant passes through a maximum at $\phi = 0.5$ [15]. This assumption allows two important further procedures. Firstly, the value of I_{110} when Q' is a maximum (I_{Q*}) may be interpreted as being that where the volume fraction of crystals $\phi = 0.5$. The relationship between volume and mass fractions of crystals is simply

$$X_c = \phi \rho_c / \rho \qquad \{8\}$$

where ρ_c is the density of pure crystal and ρ is the density of the semicrystalline polymer. Therefore two points on the linear scale $X_c = \phi \rho_c / \rho = I_c / I_t$ are known for $\phi = 0$ and $\phi = 0.5$ in the limit that the semi-crystalline polymer density varies linearly between 0.93 at 100 °C and 0.85 at the melting point [39]. Secondly, the quadratic may be solved using the value of Q' at $\phi = 0.5$ to give the electron density at that temperature of $<\eta^2> = 328.75$ arbitrary units. The electron density difference is also temperature dependent but this is small compared with the changes in ϕ. Solution of the quadratic for Q' gives two sets of solutions in ϕ which may be compared with that calculated from I_{110} so that the most reasonable values are taken. The values of ϕ are calculated from the WAXS data by

$$\phi \text{(WAXS)} = I - I_a / 2(I_{Q*} - I_a) \qquad \{9\}$$

and values of ϕ calculated from the quadratic equation of the invariant, corrected for the linear change in the (electron) density difference [16], are plotted against time in Figure 7. The crosses are the WAXS data and the circles are the SAXS data (filled and open circles being the two sets of solutions). The correlation between the degree of crystallinity from WAXS and the degree of crystallinity from SAXS is good considering the assumptions made in the calculation.

In the two phase model the average lamella thickness, L, is given by the product of the degree of crystallinity, ϕ, and the long spacing, d_1. Figure 8 is a plot of the measured parameters ϕ and d_1 and the calculated value of L according the two phase model (which is included as an inset) [6,40,41]. The degree of crystallinity falls and the long spacing increases with temperature as one would expect; individual lamellae melt out in a first order fashion, the lower order lamellae melting at lower temperatures. The average lamella thickness remains constant prior to falling asymptotically at the melting point. This is counter-intuitive as the thinnest lamellae (those having the greatest interfacial area) should melt first causing the d-spacing to increase [6,38,40]. The experimental result can be rationalised, however, by considering a growing interface causing an apparent reduction in ϕ.

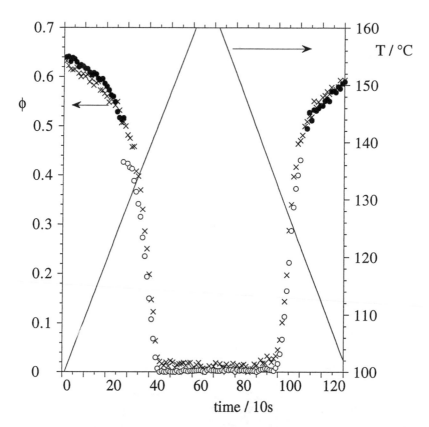

Figure 7. The degree of crystallinity, ϕ, calculated from the integrated WAXS intensity (x) and from the SAXS invariant (o and ●) and the temperature (—) as a function of time during melting and recrystallisation of HDPE.. The ϕ from WAXS is calibrated from the integrated intensity across the 110 peak associated maximum in Q' being 50% crystallinity [15] and its value at T = 160 °C being zero crystallinity. The ϕ from SAXS is calculated by solving the quadratic in Q' and correcting for changes in density.

III   SAXS/WAXS of a semi-crystalline polyurethane (PU).

The polyurethane samples studied had an initial microphase separated morphology which was formed by annealing at 120 °C. 256 x 6 s frames of SAXS and WAXS data were collected in the 100 -> 220 -> 160 °C thermal cycle with heating and cooling rates of 20 °C min^{-1}. The data in Figures 9 and 10 show the initial (100 °C), molten (220 °C), and final annealed (160 °C) patterns, the initial and molten SAXS and WAXS patterns have been shifted 25 and 50 units and 5 and 10 units respectively for ease of viewing. The experimental data were treated as described in the experimental section, the low intensity WAXS peaks at 31, 34 and 36° are from the sample holder. The uppermost Lorentz corrected SAXS curve in Figure 9 shows a broad peak (at q*) in $I(q)q^2$ typical of a lamella crystal structure with a broad distribution of crystal size. Application of Bragg's law gives a d_1 of 140 Å which is in agreement with previous studies of this type of material [11-13, 17-19]. The corresponding WAXS pattern (the uppermost curve in Figure 10) shows little evidence of good crystal structure with a peak which is only slightly different to the amorphous halo of the molten polymer, there is a shoulder at $2\theta = 22°$ and a weak peak at $\approx 19°$. The rate of crystallisation of the polyurethane

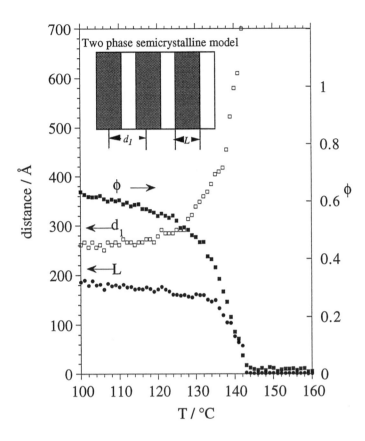

Figure 8. The degree of crystallinity, ϕ (■), and long spacing, d_I (□), with their product the lamellar thickness, L (●), plotted as a function of temperature. The relationship between the parameters is illustrated in the two phase semicrystalline model which is included as an inset.

hard segments is low at 120 °C (the temperature the material was annealed at) and generally only low degrees of crystallinity are observed by DSC [10] or WAXS [13]. The molten material still has some SAXS structure (the weak feature at $q = 0.04$ Å-1) which is either due to incomplete isotropisation or fluctuation scattering and the strong upturn at low q is typical of molten polyurethanes [11-13, 17-19], the corresponding WAXS pattern is a typical amorphous halo. The annealed material has a much sharper SAXS peak with a d_I of 170 Å, the corresponding WAXS pattern shows strong crystalline reflections at 17.5°, 19°, 21° and 23°. Thus the well-ordered crystals which grow at 160 °C (close to the maximum growth rate) give rise to a clearly defined lamella structure with a longer repeat.

The dynamics of structure development during the experiment may be observed in the contour plots in Figure 11. SAXS and WAXS contours are in arbitrary levels of $I(q)q^2$ and $I(2\theta)$ respectively. The peak position in the SAXS pattern has been marked with a bold line, q^*, and remains constant up to 120 °C after which the structure starts to change continuously by lamella thickening. This process involves melting, diffusion and some recrystallisation and is manifest as a shift in q^* to lower values of q: the weak WAXS pattern is relatively insensitive to these dynamic processes. Eventually the hard segments melt (≈ 210 °C) and mix with the soft segment, the SAXS peak

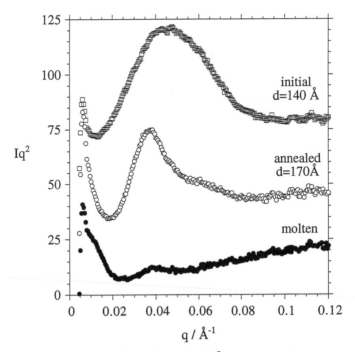

Figure 9. SAXS Lorentz corrected intensity, Iq^2, versus scattering vector, q, for the initial (□), annealed (○) and molten (●) polyurethanes.

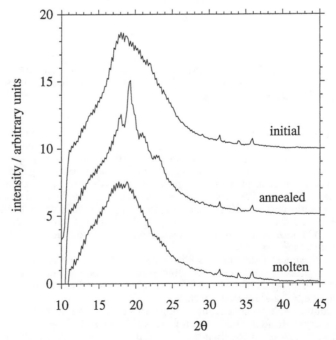

Figure 10. WAXS scattered intensity versus scattering angle, 2θ, for the PU.

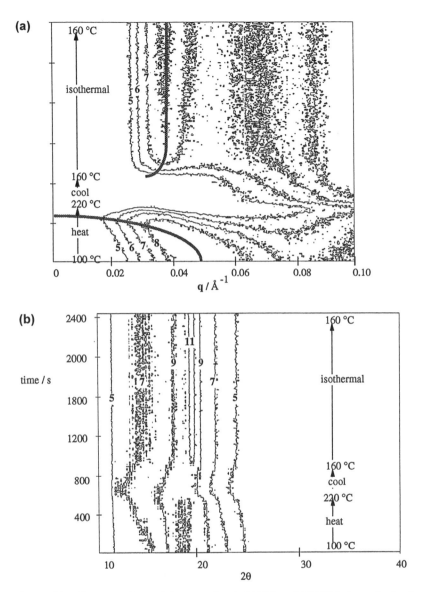

Figure 11. Contour plots of time-resolved SAXS and WAXS data obtained during melting and annealing of a copolyurethane. The SAXS data are isointensity contours in the Lorentz corected intensity, Iq^2, versus scattering vector , q, versus time. The WAXS data are isointensity contours in the scattered intensity, I, versus the scattering angle 2θ, versus time. The numbers on the contours are unrelated arbitrary levels.

disappears and the weak reflection at $2\theta = 19°$ reduces to the amorphous halo. On cooling a strong feature appears in the SAXS pattern (marked by a bold line) which rapidly comes to a constant value. There is a corresponding change in the WAXS pattern with growth in the reflections at 17.5°, 19°, 21° and 23°; the 19° reflection being the most obvious.

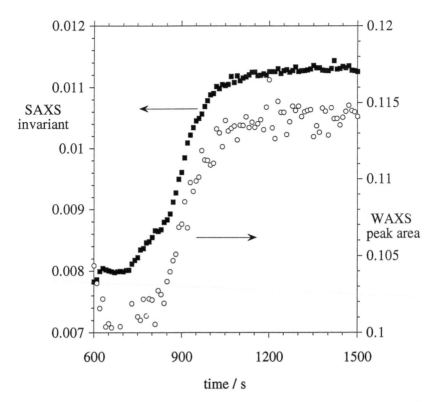

Figure 12. Evolution of structure during annealing from SAXS invariant (■) and the integrated area of WAXS peak (o).

The data may be analysed further to give some insight into the kinetics and mechanism of the structuring process. The degree of crystallinity is often assessed by measuring the ratio of the area under crystalline and amorphous WAXS patterns. The combination of weak structure and a count rate limited WAXS detector does not allow this to be done but a first order approximation to the structure development has been obtained by integrating the WAXS intensity between 18.5° and 19.5° which encompasses the strongest crystalline peak.

Figure 12 shows the evolution of structure during the annealing process. At an elapsed experimental time of 600 s the material is at 220 °C and is cooled at 20 °C min^{-1} to reach 160 °C at 780 s. Both the SAXS invariant Q' and the integrated WAXS intensity start to grow at ≈ 800 s. The development of structure has a sigmoidal shape typical of a nucleation and growth process. In fact the SAXS and WAXS structure develops with the same kinetics. Previous studies on polyurethanes which have amorphous hard segments indicate that the order-disorder transition of the block copolymer occurs at ≈ 140 °C [42]. Thus annealing this polymer at 160 °C leads to the crystallisation of the hard segment from an isotropic (disordered) block-copolymer melt and SAXS structure grows at the same rate as the WAXS structure. Previous time-resolved independent SAXS and WAXS experiments on similar materials [11-13] did not have sufficient time resolution to be able to offer such an unambiguous statement.

Subsequent off-line DSC experiments show the as-prepared polyurethane to have a hard segment melting point of 178 °C and a heat of fusion of 20J/g whereas the annealed polyurethane had a hard segment melting point of 198 °C and a heat of fusion of 30 J/g. The off-line DSC experiments confirm that a weak structure formed at 120 °C was transformed into a more ordered structure by annealing at 160 °C.

Summary and Conclusions

SAXS/DSC has been shown to be a viable technique previously [11-13] and the new SAXS/DSC at the SRS has been used to follow the recrystallisation and melting of PET. The agreement in the location of transitions between the differential of the invariant with respect to temperature and the DSC-curve is within the time-resolution of the SAXS instrument.

The SAXS/WAXS camera has been used to assess the degree of crystallinity and lamella thickness of HDPE from simultaneously obtained SAXS and WAXS patterns during melting and crystallisation. The long spacing was obtained, in the conventional manner, from the peak in Lorentz corrected SAXS pattern. The degree of crystallinity was obtained by a combination of the SAXS invariant and the integrated WAXS intensity due to the crystals. The invariant passes through a maximum at 50 % crystallinity [41,16] and the integrated WAXS intensity is a minimum for the melt. The maximum in Q' calibrates the corresponding I_{110} to be that of $\phi = 0.5$, the melt value of I_{110} corresponds to $\phi = 0$: two points in the linear relationship between WAXS intensity and the degree of crystallinity are known. Calculation of the crystallinity by solving the quadratic of ϕ in Q' is in good agreement with the WAXS once the changes in density have been taken into account. The long spacing and the degree of crystallinity may be combined in the two-phase model to estimate the lamella thickness. The melting process is characterised by a gradual reduction in the degree of crystallinity and an increase in the long spacing, this is because of first-order melting of low order crystals. These two effects combine to give a lamella spacing that does not appear to change during the melting process which is counter-intuitive [6].

The experiments on polyurethane illustrate the utility of the SAXS/WAXS camera in differentiating between microphase separation and crystallisation in semi-crystalline block copolymers.

New experiments on the in-situ growth of zeolites, annealing of cordierite glasses and phase transformations on a wide variety of polymer systems have been performed in both the SAXS/DSC and the SAXS/WAXS mode. The combination of SAXS, WAXS and DSC has been recently developed and will be reported elsewhere. A combined FTIR experiments with the X-ray scattering techniques have recently been done and this technique combination also hold great promise.

Acknowledgments
These experiments would have been impossible without the support of Neville Greaves, Jim Sheldon, Dave Bouch, Paul Hindley and Brian Parker.

Literature Cited
1. MacKenzie, R.C.; Laye, P.J.; *Chemistry in Britain*, **1986**, 22, 1005.
2. Obalisi, O.; Robeson, R.M.; Shaw, M.T. *Polymer-Polymer Miscibility*, Academic Press, New York 1979.
3. Ryan, A. J.; Stanford, J.L.; Still, R.H.; *British Polymer Journal*, **1988**, 20 ,77.
4. Camberlin, Y.; Pascault, J.P.; *J. Polym. Sci. (Polym. Chem. Edn.)*, **1983**, 27, 415
5. Cullity, D.B. *Elements of X-ray Diffraction*, Addison Welesly: Reading, 1978.
6. Vonk, C.; Baltá-Calleja, F.J., *X-ray Scattering of Polymers*, Elsevier, Amsterdam 1989.
7. Charlsley, E. L.; Warrington, S. B.; *Thermal Analysis-Techniques and Applications*, RSC,London, 1992

8. Catlow, C. R. A and Greaves, G. N., *Applications of Synchrotron Radiation*, Blackie, Glasgow 1980.
9. Koberstein, J.T.; Russell, T.P.; *J. Polym. Sci., Polym. Phys. Ed.*, **1985**, *23*, 1109.
10. Leung, L.M.; Koberstein, J.T.; *J. Polym. Sci., Polym. Phys. Ed.*, **1985**, *23*, 1883.
11. Koberstein, J.T.; Russell, T.P.; *Macromolecules*, **1986**, *19*, 714
12. Koberstein, J.T.; Yu, C.C.; Galambos, A.F.; Russell, T.P.; Ryan, A.J.; *Polym. Prep. Am. Chem. Soc. Div. Polym. Chem.*, **1990**, *31 (2)*, 110.
13. Koberstein, J.T.; Galambos, A.F.; *Macromolecules*, **1992** 25 5618.
14. Ungar, G; Feijoo, J.L.; *Mol. Cryst. Liq. Cryst.*, **1990**, *180B*, 281.
15. Gehrke, R.; Reikel, C.; Zachmann, H. G.; *POLYMER*, **1989**, *30*, 1582.
16. Schoeterden, P.; Vandermarliere, M.; Reikel, C.; Koch, M. H. J.; Groeninckx, G.; Reynaers, H.; *Macromolecules*, **1989** 22 237.
17. Li, Y.; Gao, T.; Chu, B.;*Macromolecules*, **1992**, *25*, 1737.
18. Chu, B.; Gao, T.; Li, Y.; Wang, J.; Desper, C. R.; Byrne, C. ;*Macromolecules*, **1992**, *25*, 5724.
19. Li, Y.; Ren, Z.; Zhao, M.; Yang, H.; Chu, B.;*Macromolecules*, **1993**, *26*, 612.
20. Chu, B.; Wu, D. Q.;Lundberg, R. D.; MacKnight, W. J.*Macromolecules*, **1993**, *26*, 994.
21. Wang, J.; Li, Y.; Peiffer, D. G.; Chu, B.;*Macromolecules*, **1993**, *26*, 2633.
22. Cogan, K. A.; Gast, A. P.; Capel M.; *Macromolecules* **1991**, *24*, 2633.
23. Bark, M.; Schulze, C.; Zachmann, H.G.; *Polym. Prep. Am. Chem. Soc. Div. Polym. Chem.*, **1990**, *31(2)*, 106.
24. Kruger, K. N.; Zachmann, H.G.; *Macromolecules*, **1993**, *26*, 5202
25. Tashiro, K.;Satkowski, M. M.; Stein, R. S.; Li, Y.; Chu, B.; Hsu, S. L.;*Macromolecules*, **1992**, *25*, 1809.
26. Hsiao, B.S.; Gardener, K. G.; Wu, D. Q.; Chu, B.;*Polymer*, **1993**, *34*, 3987.
27. Hsiao, B.S.; Gardener, K. G.; Wu, D. Q.; Chu, B.;*Polymer*, **1993**, *34*, 3998.
28. Bras, W.; Derbyshire, G. E.;Ryan, A,. J.; Mant, G. R.; Greaves, G. N.;*Inst. Phys. Conf. Ser., 1992, 180*, 635.
29. Bras, W.; Derbyshire, G. E.;Ryan, A,. J.; Mant, G. R.; Felton, A.; Lewis, R. A.; Hall, C. J.; Greaves, G. N.;*NIMPR* A **1993**, *A326*, 587.
30. Ryan, A. J.; *Journal of Thermal Analysis*, **1993**, 40, 887.
31. Ryan, A. A. J.; Bras, W.; Derbyshire, G. E.;Mant, G. R.; *Polymer*, **1994**, *35*, in press.
32. O'Kane, W.;Young, R. J.; Ryan, A. J.; Bras, W.; Derbyshire, G. E.;Mant, G. R.; *Polymer*, **1994**, *35*, 1352.
33. Ryan, A. J.; Naylor, S.;Bras, W.; Derbyshire, G. E.;Mant, G. R.; *Polym. Mat. Sci. Eng.*, **1993**, *69*, 449.
34. Jenkins, P.; Cameron, R. E.; Donald A. M.; Ryan, A. J.; Bras, W.; Derbyshire, G. E.;Mant, G. R.; *J. Polym. Sci. Polym. Phys. Edn.*, **1994**, *32*, 1579.
35. Lewis R.A.; Sumner, I; Berry, A.; Bordas, J.; Gabriel, A; Mant, G.; Parker, B.; Roberts, K.; Worgan, J; , NIMPRA, **1988**,*A273*, 773.
36. Bras, W.; Derbyshire, G. E.; Mant, G. R.; Clarke, S.; Cooke, J.; Komanschek, B.; Ryan, A,. J.;*J. Appl. Cryst.* **1994** *in press*, Daresbury Preprint DL/SCI/P909E
37. Porod, G.; Kolloid Z-Z, **1951** *124* 83.
38. Young, R. J.; Lovell, P. A.; *Introduction to Polymers*, Chapman and Hall, London, 1991.
39. van Krevelen, D. W.; *Properties of Polymers: Correlations with Chemical Structure*, 3rd. ed, Elsevier: Amsterdam, The Netherlands (1991).
40 Vonk, C. G.;*J.Appl. Cryst.*, *197 8* 340.
41 Strobl, G. R.; Schnieder, M. J.; *J. Polym. Sci., Polym. Phys. Ed.*, **1981** *19* 1361.
42 Ryan, A.J.; Bras, W.; Macosko, C. W.; *Macromolecules*, **1992**, *25*, 6277.

RECEIVED August 25, 1994

INDEXES

Author Index

Affiliation Index

Subject Index

Bestsellers from ACS Books

The ACS Style Guide: A Manual for Authors and Editors
Edited by Janet S. Dodd
264 pp; clothbound ISBN 0–8412–0917–0; paperback ISBN 0–8412–0943–X

The Basics of Technical Communicating
By B. Edward Cain
ACS Professional Reference Book; 198 pp;
clothbound ISBN 0–8412–1451–4; paperback ISBN 0–8412–1452–2

Chemical Activities (student and teacher editions)
By Christie L. Borgford and Lee R. Summerlin
330 pp; spiralbound ISBN 0–8412–1417–4; teacher ed. ISBN 0–8412–1416–6

Chemical Demonstrations: A Sourcebook for Teachers,
Volumes 1 and 2, Second Edition
Volume 1 by Lee R. Summerlin and James L. Ealy, Jr.;
Vol. 1, 198 pp; spiralbound ISBN 0–8412–1481–6;
Volume 2 by Lee R. Summerlin, Christie L. Borgford, and Julie B. Ealy
Vol. 2, 234 pp; spiralbound ISBN 0–8412–1535–9

Chemistry and Crime: From Sherlock Holmes to Today's Courtroom
Edited by Samuel M. Gerber
135 pp; clothbound ISBN 0–8412–0784–4; paperback ISBN 0–8412–0785–2

Writing the Laboratory Notebook
By Howard M. Kanare
145 pp; clothbound ISBN 0–8412–0906–5; paperback ISBN 0–8412–0933–2

Developing a Chemical Hygiene Plan
By Jay A. Young, Warren K. Kingsley, and George H. Wahl, Jr.
paperback ISBN 0–8412–1876–5

Introduction to Microwave Sample Preparation: Theory and Practice
Edited by H. M. Kingston and Lois B. Jassie
263 pp; clothbound ISBN 0–8412–1450–6

Principles of Environmental Sampling
Edited by Lawrence H. Keith
ACS Professional Reference Book; 458 pp;
clothbound ISBN 0–8412–1173–6; paperback ISBN 0–8412–1437–9

Biotechnology and Materials Science: Chemistry for the Future
Edited by Mary L. Good (Jacqueline K. Barton, Associate Editor)
135 pp; clothbound ISBN 0–8412–1472–7; paperback ISBN 0–8412–1473–5

For further information and a free catalog of ACS books, contact:
American Chemical Society
Distribution Office, Department 225
1155 16th Street, NW, Washington, DC 20036
Telephone 800–227–5558